"""
LABORATORY MANUAL

TO ACCOMPANY

CHEMISTRY

AN INTRODUCTION TO GENERAL, ORGANIC, AND BIOLOGICAL CHEMISTRY

LABORATORY MANUAL

TO ACCOMPANY

CHEMISTRY

AN INTRODUCTION TO GENERAL, ORGANIC, AND BIOLOGICAL CHEMISTRY

Fifth Edition

KAREN C. TIMBERLAKE

Los Angeles Valley College

HarperCollins*Publishers*

Laboratory Manual to accompany *CHEMISTRY: AN INTRODUCTION TO GENERAL, ORGANIC, AND BIOLOGICAL CHEMISTRY,* Fifth Edition

Copyright © 1992 by HarperCollins Publishers Inc.

All rights reserved. Printed in the United States of America. No part of this book may be used or reproduced in any manner whatsoever without written permission, with the following exception: testing materials may be copied for classroom testing. For information, address HarperCollins Publishers Inc., 10 E. 53rd St., New York, NY 10022.

ISBN: 0-06-046578-6

95 96 9

TABLE OF CONTENTS

The Chemistry Laboratory	ix
Laboratory Equipment	x
Preparation for Laboratory Work	xii
Laboratory Safety	xii
Safety Quiz	xiv

Experiment 1:	Math and the Calculator	1
Experiment 2:	Measuring Length	11
Experiment 3:	Measuring Volume	19
Experiment 4:	Measuring Mass	27
Experiment 5:	Density and Specific Gravity	37
Experiment 6:	Temperature and Changes of State	45
Experiment 7:	Measuring Heat of Reaction	57
Experiment 8:	Measuring Heat of Fusion and Vaporization	67
Experiment 9:	Atoms and Elements	75
Experiment 10:	Periodic Properties and Electron Arrangement	83
Experiment 11:	Bonding of Elements in Compounds	93
Experiment 12:	Writing Formulas and Names of Compounds	101
Experiment 13:	Moles and Chemical Formulas	111
Experiment 14:	Chemical and Physical Changes	119

v

Experiment 15:	Formula of a Hydrated Salt	129
Experiment 16:	Detecting Radiation	137
Experiment 17:	Boyle's Law	145
Experiment 18:	Charles' Law	153
Experiment 19:	Partial Pressures of the Gases in Atmospheric and Exhaled Air	163
Experiment 20:	Characteristics of Solutions	173
Experiment 21:	Concentrations of Solutions	181
Experiment 22:	Solutions, Colloids, and Suspensions	189
Experiment 23:	Electrolytes and Insoluble Salts	197
Experiment 24:	Testing for Cations and Anions	205
Experiment 25:	Acids, Bases, pH, and Buffers	217
Experiment 26:	Acid-Base Titrations	225
Experiment 27:	Comparing Properties of Organic and Inorganic Compounds	237
Experiment 28:	Saturated Hydrocarbons	241
Experiment 29:	Some Properties of Hydrocarbons	253
Experiment 30:	Properties of Alcohols, Aldehydes, and Ketones	261
Experiment 31:	Carboxylic Acids and Esters	273
Experiment 32:	Preparation of Aspirin	280

Experiment 33:	Amines and Amides	289
Experiment 34:	Carbohydrates	297
Experiment 35:	Saponification of Lipids	309
Experiment 36:	Amino Acids	319
Experiment 37:	Proteins	327
Experiment 38:	Enzyme Activity	337
Experiment 39:	Vitamins	345
Experiment 40:	Digestion of Foodstuffs	353
Experiment 41:	Chemical Compounds in Foods and Drugs	361
Experiment 42:	Chemistry of Urine	367
Experiment 43:	Energy Production in the Living Cell	375
Appendix A:	Constructing a Graph	383
Appendix B:	Special Solutions for the Laboratory	386

THE CHEMISTRY LABORATORY

Here you are in a chemistry laboratory with your laboratory book in front of you. Perhaps you have already been assigned a laboratory drawer, full of glassware and equipment you have never seen before. Looking around the laboratory, you may see bottles of chemical compounds, balances, burners and other equipment that you are going to use. This may very well be your first experiment with experimental procedures. At this point you may have some trepidation about what is expected of you. This laboratory manual is written with those considerations in mind.

USING THE LABORATORY MANUAL

Each experiment begins with objectives. These tell you the concepts you will be studying in that experiment. Each experiment is keyed to the concepts in the text for your study review. You will also find a list of the materials needed for each particular experiment.

The procedures are written to guide you through the experiment. When you are ready to begin your experiment, remove the laboratory record sheet. As you follow the procedures for the experimental activities, record your data and answer questions on the laboratory report sheet. Complete the required calculations and answer follow-up question designed to test your understanding of the concepts from the experiment.

PURPOSE OF THE LABORATORY

It is important to realize that the value of the laboratory experience depends on the time and effort you invest in it. Only then will you find that the laboratory can be a valuable learning experience and an integral part of the chemistry class. The laboratory gives you an opportunity to go beyond the lectures and words in your textbook and experience the scientific process of experimental trial and error from which conclusions and theories concerning chemical behavior are drawn. In some experiments, the concepts are correlated with health and biological concepts. Chemistry is not an inanimate science, but one that helps us to understand the behavior of living systems.

It is my hope that the laboratory experience will help illuminate the concepts you are learning in the classroom. The experimental process can help make chemistry a real and exciting part of your life and provide you will skills necessary for your future.

LABORATORY EQUIPMENT

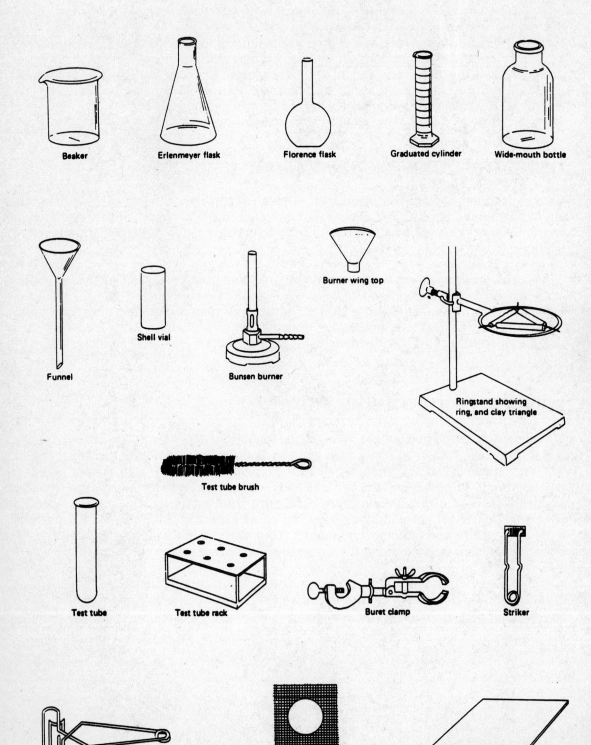

LABORATORY EQUIPMENT

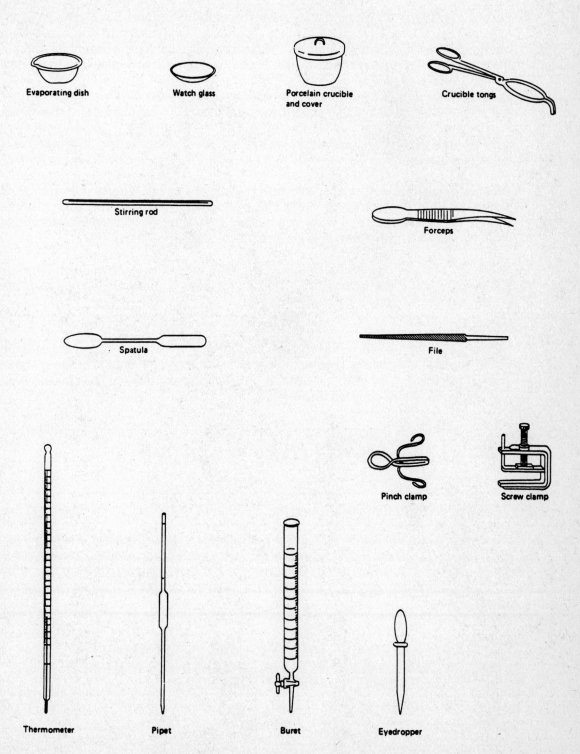

PREPARATION FOR LABORATORY WORK

1. Before you begin an experiment, read the discussion and directions for that experiment. If you have been given a laboratory schedule, read the experiment before you come to the laboratory. Make sure you fully understand the experiment before starting the actual work. If you have a question, ask your instructor to clarify the procedures.

2. Do only the experiments that have been assigned by your instructor. No unauthorized experiments are to be carried out in the laboratory. Experiments are done at assigned times, unless you have an open lab situation. Do not work alone in a laboratory.

3. Wear sensible clothing to the laboratory. Loose sleeves, shorts, or open-toed shoes can be dangerous. Tie long hair back to put it out of reach of a flame from a Bunsen burner and from chemicals. You may wish to wear protective clothing such as a lab apron or a lab coat.

4. No food, drinks or smoking is allowed in the laboratory. Wash your hands before you leave the laboratory. Do not let your friends visit while you are working in the lab; have them wait outside.

5. Before you begin a lab, clear the lab bench or work area of all your personal items such as backpacks, books, sweaters, and coats. Find a storage place in the lab for them. All you will need is your laboratory manual, calculator, text and equipment from your lab drawer.

LABORATORY SAFETY

THE CHEMISTRY LABORATORY WITH ITS EQUIPMENT, GLASSWARE AND CHEMICALS HAS THE POTENTIAL FOR ACCIDENTS. IN ORDER TO AVOID DANGEROUS ACCIDENTS, OR TO MINIMIZE THEIR DAMAGE, PRECAUTIONS MUST BE TAKEN BY EACH STUDENT AND TEACHER TO INSURE THE SAFETY OF EVERYONE WORKING IN THE LABORATORY.

 APPROVED EYE PROTECTION IS REQUIRED IN THE LABORATORY AT ALL TIMES FOR ALL EXPERIMENTAL PROCEDURES

HANDLING CHEMICALS

1. To avoid contamination of the chemical reagents, *never* insert droppers, pipets or spatulas in the reagent bottles. ***DOUBLE CHECK THE LABEL*** on the bottle before you remove a chemical.

2. Pour or transfer a chemical into a small, clean container (beaker, test tube, flask, etc.) from your drawer. Take only the quantity of chemical needed for the experiment. Make sure to cover the reagent bottle with its original cover. ***Return*** a reagent bottle to its proper location in the laboratory; do not keep a reagent bottle at your desk. Label the chemicals in the container. Many containers have etched sections on them for pencil writing. Otherwise, use tape or a marking pencil.

3. Your instructor will give directions for the disposal of used chemicals. ***Never return unused chemical to their reagent bottles.*** Some liquids and water-soluble compounds may be washed down the sink with plenty of water, but check with your instructor first. Dispose of organic compounds in specially marked containers in the hoods.

4. ***NEVER TASTE A CHEMICAL!*** Never use any equipment in the drawer such as a beaker to drink from. ***DO NOT EAT OR DRINK IN THE LABORATORY!*** If you take a break from lab, wash your hands and go outside the lab to eat a snack.

5. When required to note the odor of a chemical, first take a deep breath of fresh air. Then use your hand to waft some vapors toward your nose and note the odor. Do not inhale the fumes directly.

6. ***NEVER PIPET A LIQUID BY MOUTH!*** Use a pipet bulb.

7. Do not weigh chemicals directly on a balance pan. Use a container, watch glass, or weighing paper.

8. Do not shake down a mercury or other laboratory thermometer. Laboratory thermometers respond quickly to the temperature of their environment.

9. Begin your cleanup 15 minutes before the end of the laboratory session. Clean and dry equipment. Return any borrowed equipment to the stockroom. Turn off gas and water at your work area. Make sure you leave a clear desk. Check the balance you used. Clean up any spills in the area.

HEATING CHEMICALS

1. Only glassware marked Pyrex or Kimax can be heated. Other glassware will shatter if heated. To heat solids or liquids in a test tube, hold the tube in a holder at an angle - not upright - over the flame. Move the test tube continuously as you heat the sides and bottom. Never point the open end of the test tube at anyone, or look directly into it.

2. Never heat a flammable liquid over an open flame.

3. Be careful of picking up equipment you may just have heated. This might be an iron ring, a clay triangle, a test tube or beaker, or a crucible. A hot piece of iron or glass looks the same as it does as room temperature.

4. Do not place hot objects on a balance. Let the object cool first.

EMERGENCY PROCEDURES

1. Learn the location and use of the emergency eye-wash fountain, the emergency shower, fire blanket, fire extinguishers, and exits. Memorize their locations in the laboratory.

2. Clean up spills at your work area or floor immediately. An absorbent compound may be available to soak up the chemical. Broken glassware can be swept up with a brush and pan. Notify your instructor of any mercury spills. Mercury cleanup requires special attention.

3. If a chemical spills on the skin, **_IMMEDIATELY FLOOD THE AREA WITH WATER. CONTINUE RINSING WITH WATER FOR TEN MINUTES._** Any clothing soaked with a chemical must be removed. If left on the skin, the absorbed chemical will cause more damage. Do not try to neutralize an acid or base spilled on the skin.

4. If a chemical splashes into the eyes, flood the eyes with water at the eye-wash fountain. Get help for this procedure. Continue to rinse with water for at least ten minutes.

5. Notify your instructor of any chemical spill or accident in the laboratory.

6. If clothing or hair catch on fire, get the student to the shower or use the fire blanket. Cold water or ice can be applied to small burns. Do not use grease or other oil-based compound. Small fires can be extinguished by covering them with a watchglass. If a larger chemical fire is involved, use a fire extinguisher to douse the flames. _Do not direct a fire extinguisher at a person in the laboratory._ Shut off gas burners in the laboratory.

SAFETY QUIZ

1. Approved eye protection is to be worn
    a. for certain experiments.
    b. only for hazardous experiments.
    c. all the time.

2. Eating in the laboratory is
    a. not permitted.
    b. allowed at lunch time.
    c. all right if you are careful.

3. If you need to smell a chemical, you should
   a. inhale deeply over the test tube.
   b. take a breath of air and fan the vapors toward you.
   c. put some of the chemical in your hand, and smell it.

4. When heating liquids in a test tube, you should
   a. move the tube back and forth through the flame.
   b. look directly into the open end of the test tube to see what is happening.
   c. hold it tightly in your hand so it won't fall and break.

5. Unauthorized experiments are
   a. all right as long as they don't seem hazardous.
   b. all right as long as no one finds out.
   c. not allowed.

6. If a chemical is spilled on the you skin, you should
   a. wait to see if it stings.
   b. flood the area with lots of water for ten minutes.
   c. add another chemical to absorb it.

7. When taking liquids from a reagent bottle,
   a. clean a dropper first.
   b. pour the reagent into a small container.
   c. put back what you don't use.

8. In the laboratory, open-toed shoes and shorts
   a. are dangerous and should not be worn.
   b. are OK if the weather is hot.
   c. are all right if you wear a lab apron.

9. When is it all right to taste a chemical?
   a. Never
   b. When the chemical is not hazardous
   c. When you use a clean beaker

10. After you use a reagent bottle,
    a. keep it at your desk in case you need more.
    b. return it to its proper location.
    c. play a joke on your friends and hide it.

11. Before starting an experiment,
    a. read the entire procedure.
    b. ask your lab partner how to do the experiment.
    c. skip to the laboratory report and try to figure out what to do.

12. Working alone in the laboratory without supervision,
    a. is all right if the experiment is not too hazardous.
    b. is not allowed.
    c. is allowed if you are sure you can complete the experiment without help.

13. You should wash your hands
    a. before you leave the lab.
    b. only if they are dirty.
    c. before eating lunch in the lab.

14. Personal items (books, sweater, etc.)
    a. should be kept on your lab bench.
    b. should be stored out of the way, not on the lab bench.
    c. should be left outside.

15. When you have taken too much of a chemical, you should
    a. return the excess to the reagent bottle.
    b. store it in your lab locker for future use.
    c. discard it using proper disposal procedures.

16. In lab, you should wear
    a. sensible, protective clothing.
    b. something fashionable.
    c. shorts and loose sleeve shirts.

17. If a chemical is spilled on the table,
    a. clean it up right away.
    b. let the stockroom help clean it up.
    c. forget about it.

18. If mercury is spilled,
    a. pick it up with a dropper.
    b. call your instructor.
    c. push it under the table where no one can see it.

19. If you hair or shirt catches on fire, you should
    a. run to the nearest exit.
    b. let it burn out.
    c. use the fire blanket to put it out.

20. A liquid is taken into a pipet
    a. using a pipet bulb.
    b. using the mouth to pipet.
    c. by using a dropper.

ANSWERS TO SAFETY QUIZ

1. c	5. c	9. a	13. a	17. a
2. a	6. b	10. b	14. b	18. b
3. b	7. b	11. a	15. c	19. c
4. a	8. a	12. b	16. a	20. a

COMMITMENT TO SAFETY IN THE LABORATORY

I have read the laboratory preparation and safety procedures and agree that I will

1. Wear proper eye protection in the laboratory at all times.

2. Not perform any unauthorized experiments or work alone in the laboratory.

3. Inform the instructor of any chemical spills and/or accidents in the laboratory.

4. Not eat, drink, or smoke in the laboratory.

5. Wear sensible clothing and tie back long hair in the laboratory.

6. Carefully observe the labels on reagent bottles, remove a small amount of reagent properly, not return to the bottle, but discard it according to disposal directions.

7. Clean up spills and broken glass immediately and leave a clean work area when I leave the lab.

8. Know the locations of fire blankets, eye-wash fountains, fire extinguishers, and exits. Draw a diagram of your laboratory room and its safety features.

_____
signature

_____
laboratory class and section

_____
date

LABORATORY MANUAL

EXPERIMENT 1
MATH AND THE CALCULATOR

GOALS

1. State the correct number of significant figures in a measurement.
2. Round off a calculated answer to the correct number of significant figures.
3. Perform mathematical calculations and give final answers with the correct number of significant figures.
4. Write a number in scientific notation.

MATERIALS NEEDED

scientific calculator
pencil or pen

CONCEPTS TO REVIEW

significant figures
exact numbers
mathematical operations with significant figures
scientific (exponential) notation
calculator use

BACKGROUND DISCUSSION

In the sciences, we make many measurements such as the mass or length of an object. The values obtained from measurements are called *measured numbers*. It is important that you use these measured numbers correctly in calculations and that you report calculated answers properly.

EXPERIMENT 1

LABORATORY ACTIVITIES

A. MEASURED AND EXACT NUMBERS

Numbers obtained from measurement using some type of measuring tool are called *measured numbers*. Numbers obtained from counting or a definition are exact numbers; they do not required a measuring tool. For example, suppose you counted the people in the room and said there were 8 people. The number 8 is a counting number which makes it exact. The relationships in the metric system or the American system obtained by definition are also exact numbers. For example, the numbers in the definitions such as 100 cm in 1 meter and 12 inches in one foot are exact.

Example 1.1 Describe each of the following as a measured or exact number.
     a. 14 inches   b. 14 pencils   c. 60 minutes in 1 hour   d. 7.5 kg

Solution: a. measured
               b. exact (counting number)
               c. exact (definition)
               d. measured

Proceed to section A of the *LABORATORY REPORT* and complete the answers.

B. SIGNIFICANT FIGURES

The number of figures reported in a measured number are called *significant figures*. To count significant figures, start with the first nonzero digit. Leading zeros in decimal numbers are not significant; they are placeholders. Zeros between other digits or at the end of a decimal number are counted as significant figures. In large numbers, zeros are not significant when they express the magnitude of the number. Examples of counting significant figures in measured numbers are given in Table 1.1:

Table 1.1 Examples of Counting Significant Figures

Measured Number	Number of Significant Figures	Reason
455.2 cm	4	nonzero digits
0.80 m	2	zero follows in decimal number
50.20 l	4	zeros between and following nonzero digits
0.0005 lb	1	leading zeros are not significant
25,000 ft	2	placeholder zeros are not significant

Now complete Section B on your *LABORATORY RECORD* which can be found at the end of this experiment.

MATH AND THE CALCULATOR

C. ROUNDING OFF

After you have obtained some measurements, you may use them in mathematical operations such as multiplication, division, addition or subtraction. The results obtained from such operations are called *calculated answers*. When a calculated answer has more numbers than the measurements, it is necessary to *round off* the calculated answer. If the first number to be dropped is *less than 5*, it and all following numbers are simply dropped. However, if the number dropped is a *5 or greater*, the last retained digit is increased by 1.

Example 1.2 Round the numbers 75.6243, 0.528392, and 385,600 off to give three significant figures; two significant figures:

Solution:

		THREE SIG FIGURES	TWO SIG FIGURES
75.6243	rounds off to	75.6	76
0.528392	rounds off to	0.528	0.53
385,600	rounds off to	386,000	390,000

Note: When rounding off large numbers, be sure to replace dropped numbers with placeholder zeros to retain the magnitude of the number.

Now complete Section C in the *LABORATORY RECORD*.

D. MULTIPLICATION AND DIVISION

A calculated answer obtained from multiplication and/or division must be rounded off to the same number of significant figures as the measured number used in the calculation with the *smallest* number of significant figures.

Example 1.3: Calculate and round off your answer to give the correct number of significant figures: 3.56 x 4.9 =

Solution: On the calculator, the steps are:

ENTER KEY	DISPLAY READS	
3.56	3.56	
x	3.56	
4.9	4.9	
=	17.444	(calculator answer)
=	17	final answer

The calculated answer has too many figures. It must be rounded off to a final answer of 17 since the measurement 4.9 in our calculation has two significant figures.

Example 1.4: Solve: $\dfrac{(0.025)(4.62)}{3.44}$

Solution: On the calculator, the steps are:

ENTER KEY	DISPLAY READS	
0.025	0.025	
x	0.025	
4.62	4.62	
=	0.1155	
÷	0.1155	
3.44	3.44	
=	0.0335755	(calculator answer)
=	0.034	final answer

The calculator answer is rounded off to two significant figures to give an answer of 0.034.

Now complete Section D in the *LABORATORY RECORD*.

E. ADDITION AND SUBTRACTION

A calculated answer obtained from the addition or subtraction of measured numbers must be rounded off to give a final answer that has the same number of digits as the measured number in the calculation with the *fewest* digits after the decimal point.

Example 1.5: Add:

	42.11	
	4.056	
	30.1	(1 digit after the decimal point)
	76.266	(calculator answer)
	76.3	final answer

The final answer has been rounded off to contain 1 digit after the decimal point.

Example 1.6: Subtract:

	14.621	
	-3.39	(2 digits past the decimal point)
	10.231	(calculator answer)
	10.23	final answer

The calculator answer is rounded off to give a final answer with two digits after the decimal point.

Now complete Section E in the LABORATORY RECORD.

MATH AND THE CALCULATOR

F. SCIENTIFIC NOTATION

In scientific work, we work with small numbers such as 0.000000025 m and some very large numbers such as 4,000,000 g. It is sometimes convenient to express such large and small numbers in terms of a power of 10.

	power of ten		power of ten
0.00001	= 10^{-5}	10,000	= 10^4
0.001	= 10^{-3}	1,000	= 10^3
0.01	= 10^{-2}	100	= 10^2
0.1	= 10^{-1}	10	= 10^1

To write a number in scientific notation, use the following rules:

For numbers larger than 10:
- a. move the decimal point to the left until it follows the first digit in the number.
- b. state a power of ten that is equal to the number of places the decimal point moved to the left.

For numbers smaller than 1:
- a. move the decimal point to the right until it is placed just after the first digit in the number.
- b. state a negative power of ten that is equal to the number of places the decimal point moved to the right.

Example 1.7: Write the following values in scientific notation:
   a. 35,000   b. 0.0042   c. 0.0000815

Solution: a. 35,000 = 3.5000 = 3.5×10^4

b. 0.0042 = 0.0042 = 4.2×10^{-3}

c. 0.0000815 = 0.00008.15 = 8.15×10^{-5}

Now complete Section F in the **LABORATORY RECORD**.

EXPERIMENT 1
MATH AND THE CALCULATOR
LABORATORY REPORT

A. MEASURED AND EXACT NUMBERS

Describe each of the following numbers as a *measured* or *exact* number:

a. 5 books   _____

b. 5 lb   _____

c. 9.25 g   _____

d. 12 roses   _____

e. 12 inches in 1 foot _____

f. 361 miles   _____

g. 1000 mg in 1 g   _____

B. SIGNIFICANT FIGURES

State the number of significant figures in each of the following measured numbers:

a. 4.5 m   _____   e. 204.52 g   _____

b. 0.0004 L   _____   f. 625,000 mm   _____

c. 805 lb   _____   g. 1.0065 km   _____

d. 0.0250 L   _____   h. 82.05 g   _____

C. ROUNDING OFF

Round off each of the following numbers to the number of significant figures indicated:

	THREE SIGNIFICANT FIGURES	TWO SIGNIFICANT FIGURES
a. 49.34	_____	_____
b. 5.448	_____	_____
c. 85.83423	_____	_____
d. 532,800	_____	_____
e. 143.63212	_____	_____

D. MULTIPLICATION AND DIVISION

Perform the following multiplication and division calculations. Give a final answer with the correct number of significant figures:

a. 4.5×0.28 _____

b. $0.1184 \times 8.0 \times 0.034$ _____

c. $\dfrac{11.4}{2.3}$ _____

d. $\dfrac{(42.4)(5.6)}{1.5}$ _____

e. $\dfrac{(35.56)(1.45)}{(4.8)(0.56)}$ _____

E. ADDITION AND SUBTRACTION

Perform the following addition and subtraction calculations. Give a final answer with the correct number of significant figures.

a. 6.25 g + 0.683 g _____

b. 13.45 mL + 6.5 mL + 0.4552 mL _____

c. 145.5 m + 86.58 m + 1045 m _____

d. 245.625 kg - 80.2 kg _____

e. 4.62 cm - 0.885 cm _____

F. SCIENTIFIC NOTATION

Write the following numbers in scientific notation:

a. 4,450,000 _____

b. 38,000 _____

c. 25.2 _____

d. 0.00032 _____

e. 0.0000000021 _____

Write the following as ordinary numbers:

a. 4×10^2 _____

b. 5×10^4 _____

c. 1.88×10^6 _____

d. 8×10^{-3} _____

e. 4.25×10^{-5} _____

EXPERIMENT 1

QUESTIONS AND PROBLEMS

1. A number that counts something is an exact or pure number. When you say 4 cups, or 5 books or 2 watches, you are using exact numbers. However, when you use a ruler to measure the height of your friend as 155.2 cm, you obtain a measured number. Why are some numbers called exact numbers while other numbers are called measured numbers?

2. Which type of numbers (exact or measured) require that you count the number of significant figures for calculations?

2. Bill and Beverly have measured the sides of a rectangle. Each recorded the length of the rectangle as 6.7 cm and the width as 3.9 cm. When Bill calculates the area (by multiplying the length by the width), he gives an answer of 26.13 cm^2. However, Beverly records her answer as 26 cm^2 for the area.

    a. Why is there a difference between the two answers when they used the same measurements?

    b. You are going to tutor Bill. What would you tell him to help him correct his answer?

EXPERIMENT 2
MEASURING LENGTH

GOALS

1. Read metric measurements for length correctly.
2. Record measurements of length correctly.
3. Calculate experimental values of metric-American conversion factors.

MATERIALS NEEDED

metric ruler
string
inch ruler

CONCEPTS TO REVIEW

metric prefixes
conversion factors
measuring length in the metric system

BACKGROUND DISCUSSION

Scientists and allied health personnel must be able to understand and carry out laboratory procedures, take measurements, read thermometers, and report results accurately and clearly. How well they do these things can mean life or death to a patient.

The system of measurement used in science, and increasingly used in hospitals and clinics, is the metric system, which is a *decimal system* (based on units of 10). The unit of length in the metric system is the *meter (m)*. Using an appropriate prefix, you can indicate a length that is greater or less than a meter.

A **meterstick** *is divided into 100 centimeters*. The numbers along the meterstick are centimeters. See Figure 2.1. Each centimeter is divided into ten millimeters. Therefore, a meterstick can be read accurately to one-tenth of a centimeter. Table 2.1 lists some of the most commonly used metric units of length.

EXPERIMENT 2

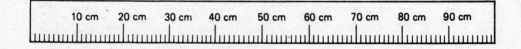

FIGURE 2.1 Markings in centimeters on a meterstick.

Table 2.1 Some Metric Units Used to Measure Length

1 kilometer (km)	= 1000 meters (m)	
1 decimeter (dm)	= 0.1 m (1/10 m)	1 m = 10 dm
1 centimeter (cm)	= 0.01 m (1/100 m)	1 m = 100 cm
1 millimeter (mm)	= 0.001 m (1/1000 m)	1 m = 1000 mm

Conversion Factors

A measurement can be expressed in American units as well as in the metric system. If we measure the same quantity in two different units, we have an *equality*. For example, we know that 1 hour is the same amount of time as 60 minutes. This relationship is called an *equality*.

$$\text{Equality} \qquad 1 \text{ hr} = 60 \text{ min}$$

When the equality is written in the form of a fraction, the relationship is called a *conversion factor*. Two fractions are possible and both are conversion factors for the equality.

Conversion Factors for 1 hr = 60 min

$$\frac{60 \text{ min}}{1 \text{ hr}} \quad \text{and} \quad \frac{1 \text{ hr}}{60 \text{ min}}$$

MEASURING LENGTH

LABORATORY ACTIVITIES

 BE SURE TO PUT ON YOUR SAFETY GOGGLES BEFORE YOU BEGIN THIS EXPERIMENT!

A. THE METERSTICK

Obtain a meterstick or ruler. Observe the markings on the meterstick. Answer the questions for part A. in the laboratory record.

B. ESTIMATIONS AND MEASUREMENTS

1. **Estimating length**. Now try some estimations using metric units. An estimation is a guess about the length of something **without measuring**. Make an estimation (your best guess) in centimeters (cm) for the length of your little finger, the distance from your elbow to your wrist, the distance around your wrist, the length of your foot, and the distance around your head.

2. **Measuring length**. When you have completed your estimations, use the meterstick to measure the same lengths in centimeters (cm). **Read the meterstick to the tenths place.** For example, if the end of the measurement is three millimeters past the 18 centimeter mark, record the length as 18.3 cm.

    Measure as carefully as you can. String may be used to obtain the distance around your head and wrist. Record the actual measurements next to the estimations you made. *Note: Do not round off any numbers obtained from measurement. They are all significant figures*.

C. COMPARING CENTIMETERS AND INCHES

In this section, we will measure a distance in the metric system using centimeters (cm), and then in the American system using inches. Figure 2.2 shows a comparison of a metric ruler and a ruler in inches.

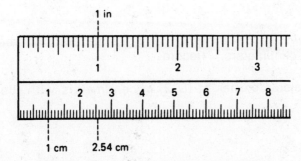

FIGURE 2.2 *A comparison of inches and centimeters.*

13

EXPERIMENT 2

C.1 Using a metric ruler, measure the height (vertical distance) of this page in cm. Record your measured value. Then using a ruler with inches, measure the same page height again using inches. Record.

Note: Do not mix units in measurements. Express a fraction of an inch or centimeter as a decimal. For example, a measurement of 4 inches and 1/2 inch is recorded as 4.5 inches. A metric measurement of 8 cm and 4 mm is recorded as 8.4 cm.

MEASUREMENT	DECIMAL FORM
4 1/2 in.	4.5 in.
10 3/4 in.	10.75 in.
8 cm and 4 mm	8.4 cm

C.2 Now you are ready to calculate a conversion factor using your measurements. Write a fraction by placing the height in cm in the numerator and the height in inches in the denominator. To simplify the fraction, *divide the numerator (cm) by the number of inches in the denominator.* Be sure to round off your answer to the correct number of significant figures. Your answer is your <u>experimental</u> value for the number of centimeters in 1 inch.

$$\frac{\text{length (cm)}}{\text{length (in.)}} = \frac{\text{cm}}{1 \text{ inch}}$$

C.3 Compare your experimental conversion factor (cm/in) with the given value of 2.54 cm/in.

D. MEASURING YOUR HEIGHT IN CENTIMETERS

D.1 *Measuring your height in inches.* Using a yardstick or an inch ruler, measure your height (with shoes) in inches. Record.

D.2 *Calculating your height in centimeters.* Using the value for your height (D.1), use the conversion factor (cm/inch) to *calculate* your height in centimeters. In the space left for calculations, write the proper setup for the calculation.

$$\text{measured height (inches)} \times \frac{2.54 \text{ cm}}{1 \text{ inch}} = \text{height (cm)}$$

D.3 *Measuring your height in centimeters.* Using a meterstick, measure your height (with shoes) in centimeters. Record.

D.4 Now compare your calculated height (D.2) in centimeters with the height you measured (D.3).

14

NAME_____ SECTION_____ DATE_____

EXPERIMENT 2
MEASURING LENGTH
LABORATORY REPORT

A. THE METERSTICK

1. What are the units represented by the <u>numbers</u> on the meterstick?

2. What do the smallest units marked on the meterstick represent?

3. Complete the following statements:

    There are __100__ centimeters(cm) in 1 meter(m).

    There are __1000__ millimeters(mm) in 1 meter(m).

    There are _____ millimeters(mm) in 1 centimeter(cm).

B. ESTIMATIONS AND MEASUREMENTS

Item	Estimation of length	Measurement of length
little finger	_____ cm	_____ cm
elbow to wrist	_____ cm	_____ cm
around wrist	_____ cm	_____ cm
foot	_____ cm	_____ cm
around head	_____ cm	_____ cm

EXPERIMENT 2

C. COMPARING CENTIMETERS AND INCHES

C.1   Height of page   _____ cm

       Height of page   _____ inches

C.2   Comparison of centimeters and inches

$$\frac{\underline{\hspace{1cm}}\text{cm}}{\text{inch}} = \frac{\underline{\hspace{1cm}}\text{cm}}{1 \text{ inch}}$$

C.3   How does your experimental value compare to the known conversion factor of 2.54 cm/inch?

D. MEASURING YOUR HEIGHT IN CENTIMETERS

D.1   Height with shoes (measured)   _____ inches

D.2   Height (calculated)   _____ cm

       Calculation setup:

D.3   Height with shoes (<u>measured</u>)   _____ cm

D.4   How does your *measured* height in centimeters compare to the height you *calculated* in cm?

NAME_____ SECTION_____ DATE_____

QUESTIONS AND PROBLEMS

1. Give one advantage of using the metric system.

2. Measure the following lengths on the line in centimeters. Convert the measurements to the other units listed:

   A   B   C
   |_____|_____|

A-B	__26__ mm	__2.6__ cm	__.026__ m
B-C	__100__ mm	__10.__ cm	__.10__ m
A-C	__126__ mm	__12.6__ cm	__.126__ m

THE FOLLOWING PROBLEMS REQUIRE THE USE OF CONVERSION FACTORS. SHOW A SETUP FOR EACH PROBLEM. ALL NUMBERS MUST HAVE UNITS, AND UNITS MUST CANCEL. GIVE YOUR FINAL ANSWERS WITH THE CORRECT NUMBER OF SIGNIFICANT FIGURES.

3. A pencil has a length of 8.5 cm. What is its length in millimeters(mm)?

_____ mm

EXPERIMENT 2

QUESTIONS AND PROBLEMS

4. An infant has a length of 20.2 inches. What is the length of the baby in cm?

_____cm

5. A roll of tape measures 32.9 m. What is the length of the tape in inches?

_____inches

6. A fish tank required 275 cm of plastic tubing. If plastic tubing sells for 35 cents/ft, what is the cost($) to purchase the plastic tubing?

$_____

EXPERIMENT 3
MEASURING VOLUME

GOALS

1. Read and record volume measurements accurately.
2. Calculate metric-American conversion factors for volume.
3. Determine the volume of a solid by direct measurement and volume displacement.

MATERIALS

display of cylinders with liquid
test tubes, small and large
50-mL graduated cylinder
1-L graduated cylinder
measuring cup
a wood or metal solid with a regular shape
metric ruler
graduated cylinders (100, 250, or 500-mL) to fit solid

CONCEPTS TO REVIEW

metric prefixes
volume units in the metric system
volume conversion factors

BACKGROUND DISCUSSION

The volume of a substance measures the space it occupies. In the metric system, the unit for volume is the *liter*. Prefixes are used to express smaller volumes such as deciliters(dL), or milliliters (mL). One cubic centimeter (cm^3 or cc) is equal to 1 mL. The terms are used interchangeably.

liter (L)	1 L	
deciliter (dL)	0.1 L	1 L = 10 dL
milliliter (mL)	0.001 L	1 L = 1000 mL

1 cubic centimeter = cm^3 = 1 cc = 1 mL

EXPERIMENT 3

LABORATORY ACTIVITIES

 BE SURE YOU PUT ON YOUR SAFETY GOGGLES BEFORE YOU BEGIN THIS EXPERIMENT!

A. MEASURING THE VOLUME OF A LIQUID

Reading a Graduated Cylinder In the laboratory, the volume of a liquid is measured in a graduated cylinder. To read the volume of water in a graduated cylinder properly, set the cylinder on a level surface and bring your eyes even with the liquid level. You may notice that the water level is not a straight line, but curves downward to the center. This curve, called a ***meniscus***, is read at its lowest point (center) to obtain the volume measurement for the liquid. See Figure 3.1.

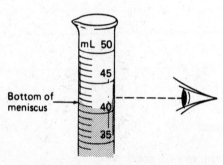

Figure 3.1 *In a graduated cylinder, the volume of the liquid, read at the bottom of the meniscus, is 42.0 mL.*

A.1 The lines or divisions on a graduated cylinder represent a specific volume. On a 50-mL graduated cylinder, each line measures 1 mL. If you estimate between the smallest lines, you can report a volume to the tenths (0.1) of a milliliter. The lines for volume on a 250-mL cylinder usually represent 5 mL. On a large graduated cylinder such as a 1000 mL cylinder, each line would be 10 mL.

Observe the liquids in a display of graduated cylinders. Record the volume (mL) of each of the liquids.

A.2 Obtain a small test tube from your laboratory equipment. Completely fill with water and transfer to a graduated cylinder. Record its volume (mL). Repeat using a large test tube.

MEASURING VOLUME

B. QUARTS AND MILLILITERS

In this section, we will determine the number of milliliters in the American volume of 1 quart. From the comparison, we can derive a metric-American conversion factor.

B.1 Measure one quart of water and transfer to a large (1-L) graduated cylinder. Record its volume in milliliter (mL). (You may also use a 1-cup measure cup four times, or a 1-pint measure cup two times.)

B.2 Write a fraction that expresses the number of milliliters in 1 quart.

B.3 Compare your experimental value with the known factor, 946 mL/1qt.

C. MEASURING THE VOLUME OF A SOLID

Direct Measurement. If an object has a regular shape such as a cube or a cylinder, its volume can be determined by measuring its dimensions.

C.1 Obtain a wood or metal solid that has a regular shape such as a cube, rectangular solid, or cylinder. Record its shape.

List the type of measurements (length, width, height, and/or diameter) you need to calculate the volume of the solid. Use a metric ruler to determine the dimensions of the solid in centimeters (cm). **Keep this object for part C.3.**

C.2 Calculate the volume of the solid in cm^3. Show your calculation using the proper formula for the volume of the solid. Include the units and round off to give an answer with the correct number of significant figures.

Shape	Dimensions	Volume
cube	length (L)	$V = L^3$
rectangular solid	length (L) width (W) height (H)	$V = L \times W \times H$
cylinder	diameter (D) height (H)	$V = \dfrac{\pi D^2 H}{4} = \dfrac{3.14 \, D^2 H}{4}$ $= \pi r^2 H$

Volume Displacement An object placed under water will cause the water level to rise. The difference in the water level before and after the object is submerged is due to the volume of the object. The object has displaced its own volume of water. By measuring the water level before and after the object is added, the volume of the object can be calculated. (See Figure 3.2.)

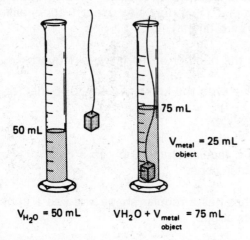

FIGURE 3.2 *Measuring the Volume of a Solid by Volume Displacement*

C.3 Obtain a graduated cylinder that is large enough to hold the solid you used in C.1. Place water in the graduated cylinder until it is about half-full. Record the level of the water.

Carefully place the solid from C.1 in the cylinder making sure it is completely submerged. Heavy solids should be tied to a piece of thread and carefully lowered to the bottom of the cylinder to prevent cracking the glass. (If a solid floats, push it under the surface of the water until it is just submerged.) Record the level of the solid and water.

Calculate the volume (mL) displaced by the solid by subtracting the initial water level from the final water level.

C.4 Convert the volume of the solid from milliliters to cubic centimeters using the equality 1 mL = 1 cm^3.

$$\text{volume of solid (mL)} \quad \times \quad \frac{1 \text{ cm}^3}{1 \text{ mL}} = \text{volume of solid (cm}^3\text{)}$$

Compare the volume of the solid obtained by **direct** measurement with the volume obtained by **volume** displacement.

NAME_____ SECTION_____ DATE_____

EXPERIMENT 3
MEASURING VOLUME
LABORATORY REPORT

A. MEASURING THE VOLUME OF A LIQUID

A.1   Volume of liquid(1)   _____ mL

        Volume of liquid(2)   _____ mL

        Volume of liquid(3)   _____ mL

A.2   Volume of small test tube   _____ mL

        Volume of large test tube   _____ mL

B. QUARTS AND MILLILITERS

B.1   Volume of 1 quart   _____ mL

B.2   Relationship of milliliters to 1 quart $\dfrac{\text{_____ mL}}{\text{1 quart}}$

B.3   How does your experimental conversion factor for mL/qt compare with the known conversion factor of 946 mL/qt?

EXPERIMENT 3

C. MEASURING THE VOLUME OF A SOLID

Direct Measurement

C.1   Shape of Solid   _____

        Dimensions

        _____   _____ cm

        _____   _____ cm

        _____   _____ cm

C.2   Volume of Solid   _____ cm³

        Formula used for volume of the solid

        V =

        Calculations for volume of the solid

Volume Displacement

C.3   Volume of water and solid   _____ mL

        Volume of water   _____ mL

        Volume of solid   _____ mL

C.4   Volume of solid   _____ cm³

        Calculation for cm³ volume of solid:

How does the volume of the solid obtained by direct measurement (C.2) compare to its volume obtained by displacement (C.4)?

NAME_____ SECTION_____ DATE_____

QUESTIONS AND PROBLEMS

1.a. A solid object has an irregular shape. What method would you use to determine its volume?

b. The water level in a graduated cylinder is 30.0 mL. A glass marble with a volume of 11.8 mL is submerged in the water. What is the final level of water in the cylinder?

_____mL

THE FOLLOWING PROBLEMS REQUIRE THE USE OF CONVERSION FACTORS. SHOW A SETUP FOR EACH PROBLEM. ALL NUMBERS MUST HAVE UNITS, AND UNITS MUST CANCEL. GIVE YOUR FINAL ANSWERS WITH THE CORRECT NUMBER OF SIGNIFICANT FIGURES.

2. A patient has received 825 mL of fluid in one day. What is that volume in liters?

_____L

EXPERIMENT 3

3. 245 mL of a cleaning solution is mixed with 0.575 L of water. What is the final volume of the solution in deciliters?

_____ dL

4. How many quarts are in 5520 mL of spaghetti sauce?

_____ qt

5. How many pints of plasma are present in 8.25 L? (1 qt = 2 pt)

_____ pt

EXPERIMENT 4
MEASURING MASS

GOALS

1. Use a laboratory balance to determine the mass of an object or substance.
2. Use the gram as a unit of mass.
3. Determine the conversion factors for grams and pounds; kilograms and pounds.

MATERIALS NEEDED

laboratory balance
small objects to weigh
small beaker
rubber stoppers
commercial product with a label showing the weight
    and mass of the contents in metric and American units
25-mL or 50-mL graduated cylinder

CONCEPTS TO REVIEW

mass
metric prefixes
percent
conversion factors for mass and weight

BACKGROUND DISCUSSION

In the metric system, the unit of mass is the *gram (g)*. A larger unit, the kilogram (kg), is used in measuring a patient's weight in a hospital, while a smaller unit of mass, the milligram (mg) is often used in the laboratory. (See Table 4.1.)

Table 4.1 Some Metric Units used to Measure Mass

kilogram (kg)	1 kg	= 1000 g
gram (g)	1 g	= 1000 mg
milligram (mg)	1 mg	= 0.001 g

EXPERIMENT 4

The *mass* of an object indicates the amount of matter present in that object. The *weight* of an object is a measure of the attraction the Earth has for that object. Since this attraction is proportional to the mass of the object, we will use the terms mass and weight interchangeably.

LABORATORY ACTIVITIES

 BE SURE YOU PUT ON YOUR SAFETY GOGGLES BEFORE YOU BEGIN THIS EXPERIMENT!

A. THE MASS OF A SOLID

Your instructor will demonstrate the method of operation for the type of laboratory balance you will be using.

A.1   Determine the mass of several solid objects. If the items listed are not available, you may select others of your own choice. Place the object on the balance pan and record its mass in the **LABORATORY RECORD**. **Do not round off any measurements of mass**. (If you are using a triple beam balance, be sure that all of your recorded measurements include a figure in the hundredth's, 0.01 place. Be sure to put a 0 in the hundredths place when it is part of the accuracy of the measurement.)

Example 4.1:   beaker 42.18 g   *recorded to the hundredth's place*

pencil 11.60 g   *recorded to the hundredth's place*

A.2   Now that you have used the balance several times, you are ready to determine the mass of an unknown. Ask your instructor for an unknown mass. Record its code number. Record the mass of the unknown and check your answer with the instructor.

B. PERCENT BY WEIGHT

When a mixture contains different items, we can weigh them and determine their percent by weight. A percent is the parts per 100. In a mixture of two different items, the percent is obtained by dividing the mass of one of the items by the mass of the whole mixture and multiplying by 100.

Percent by weight for item 1

$$\frac{\text{g of item 1}}{\text{g item 1 + g item 2}} \times 100 = \text{\% item 1}$$

MEASURING MASS

Percent by weight for item 2

$$\frac{\text{g of item 2}}{\text{g item 1 + g item 2}} \times 100 = \% \text{ item 2}$$

B.1 Obtain a beaker that will hold 2-3 stoppers. Find the mass of the beaker, and then the mass of all the stoppers.

B.2 Place all the stoppers in the beaker and determine their **total mass**. This is equal to the mass of the whole.

B.3 Calculate the *percent by weight for the beaker*.

$$\frac{\text{mass of beaker}}{\text{mass of beaker + stoppers}} \times 100 = \% \text{ beaker}$$

Calculate the *percent by weight for the stoppers*.

$$\frac{\text{mass of stoppers}}{\text{mass of beaker + stoppers}} \times 100 = \% \text{ stoppers}$$

Calculate the sum of their percentages.

C. MASS OF A LIQUID

Weighing by difference. The mass of a liquid is found by weighing by difference. First, the mass of a dry container such as a beaker is determined. Then the liquid is placed in the beaker, and the mass of the beaker and the liquid is recorded. The mass of the liquid is found by subtracting the mass of beaker from the combined mass.

C.1 Weigh a small beaker (50- or 100-ml). Record its mass. Place 25.0 mL of water in a graduated cylinder. Transfer the water to the beaker. (Do not use the volume markings found on the sides of some beakers. These are not for accurate volume measurement.) Weigh and record the combined mass of the beaker and the water.

C.2 Calculate the mass of the water by subtracting the mass of the beaker from the total mass.

EXPERIMENT 4

D. COMPARING UNITS OF MASS AND WEIGHT

D.1 *Grams and Pounds.* Many labels on commercial products list the mass of the product in both metric and American units. Obtain a commercial product and record the mass and weight listed on the label. Do <u>not</u> weigh. The mass refers to the substance that is/was in the container.) If the weight is given in ounces, convert it to pounds. (1 lb= 16 oz)

$$\text{Number of ounces} \times \frac{1 \text{ lb}}{16 \text{ oz}} = \underline{\quad} \text{ lb}$$

D.2 Set up the relationship of grams to pounds (g/lb). Divide through by the number of pounds to obtain the number of grams in 1 pound.

$$\frac{\text{number of grams}}{\text{number of lb}} = \frac{\underline{\quad} \text{g}}{\underline{\quad} \text{lb}} = \frac{\underline{\quad} \text{g}}{1 \text{ lb}}$$

Compare your experimental value with the standard value of 454 g/lb.

D.3 *Pounds and Kilograms* Convert the mass on the product label from grams to kilograms.

$$\text{number of grams} \times \frac{1 \text{ kg}}{1000 \text{ g}} = \text{number of kilograms}$$

D.4 Set up a relationship for the number of pounds in the product (D.1) and the number of kilograms from D.3. Divide through and report the answer as lb/kg with the correct number of significant figures.

$$\frac{\text{number of lb}}{\text{number of kg}} = \frac{\underline{\quad} \text{lb}}{\underline{\quad} \text{kg}} = \frac{\underline{\quad} \text{lb}}{1 \text{ kg}}$$

NAME_____ SECTION_____ DATE_____

EXPERIMENT 4
MEASURING MASS
LABORATORY REPORT

A. MASS OF A SOLID

A.1   **Object**   **Mass**

     small beaker   _____ g

     stopper   _____ g

     evaporating dish   _____ g

     watch glass   _____ g

     test tube holder   _____ g

A.2   **Unknown**

     code #_____   _____ g

     checked by instructor   ☐

EXPERIMENT 4

B. PERCENT BY WEIGHT

B.1   *Item*   *Mass*

       beaker   _____g

       stoppers   _____g

B.2   Total mass   _____g
               (beaker + stoppers)

B.3   Percent by weight(beaker)   _____%

       Calculation Setup:

       Percent by weight(stoppers)   _____%

       Calculation Setup:

       Total of the percentages   _____%

C. <u>MASS OF A LIQUID</u>

C.1   Beaker + 25.0 mL water   _____g

       Beaker   _____g

C.2   Mass of water   _____g

       Give a reason for finding the mass of a liquid by weighing by difference.

NAME_____ SECTION_____ DATE_____

D. COMPARING UNITS OF MASS AND WEIGHT

D.1   Commercial Product   _____

      Mass given on label: _____ g

      Weight given on label: _____ oz   _____ lb

D.2

$$\frac{\text{number of grams}}{\text{number of lb}} = \frac{\underline{\qquad}\text{ g}}{\underline{\qquad}\text{ lb}} = \frac{\underline{\qquad}\text{ g}}{1 \text{ lb}}$$

How does your experimental value for g/lb compare to the known of 454 g/lb?

D.3   Mass in kilograms (from label)

      _____ g × $\dfrac{1 \text{ kg}}{1000 \text{ g}}$ = _____ kg

D.4

$$\frac{\text{number of lb}}{\text{number of kg}} = \frac{\underline{\qquad}\text{ lb}}{\underline{\qquad}\text{ kg}} = \frac{\underline{\qquad}\text{ lb}}{1 \text{ kg}}$$

How does your experimental value compare with the known factor of 2.20 lb/kg?

EXPERIMENT 4

QUESTIONS AND PROBLEMS

1. What is the total mass in grams of objects that have masses of 0.200 kg, 80.0 g, and 524 mg?

_____ g

2. A beaker has a mass of 225.08 g. When a liquid is added to the beaker, the combined mass is 278.25 g. What is the mass of the liquid alone?

_____ g

3. A sugar solution consists of 45.8 g sugar and 108.5 g water. What is the percent by weight of sugar and the percent by weight of water in the solution?

_____ % sugar

_____ % water

SOLVE THE FOLLOWING PROBLEMS USING CONVERSION FACTORS IN A PROBLEM SETUP THAT SHOWS UNIT CANCELLATION.

4. How many mg is 0.078 g?

_____ mg

NAME_____ SECTION_____ DATE_____

5. An infant has a mass of 3.40 kg. What is the weight of the infant in pounds?

_____lb

Medication Problems

6. The doctor's order is Antabuse 240 mg. The stock on hand is 0.060-g tablets. How many tablets are needed to fill the order?

_____tablets

7. In the apothecary system, there are 60 mg in a grain (gr) or 60 mg/gr. The order is Mesurin 20 gr. The stock on hand is 600-mg tablets. How many tablets are to be administered?

_____tablets

8. The order is Pen V K 500 mg. The label reads 250 mg per teaspoon (250 mg/tsp); 1 tsp equal 5 mL (5 cc). How many mL of the medicine should be given?

_____mL

EXPERIMENT 5
DENSITY AND SPECIFIC GRAVITY

GOALS

1. Calculate the density of a solid or liquid from mass and volume.
2. Calculate the specific gravity of a liquid from its density.
3. Determine the specific gravity of a liquid using a hydrometer.
4. Graph the relationship between the mass and volume of a liquid.

MATERIALS NEEDED

50-mL graduated cylinder
small beaker
liquid (isopropyl alcohol or other)
solid objects

graduated cylinders to fit solid objects
set of liquids in graduated cylinders with hydrometers
25.0 mL of a liquid for measuring mass and volume for a graph

CONCEPTS TO REVIEW

measuring mass
measuring volume
density
specific gravity
graphing data

BACKGROUND DISCUSSION

Density

The density of a substance represents a relationship between the mass of a substance and its volume. Hospital laboratories determine the density of urine as part of a health checkup. The specific gravity of your car battery fluid is checked using a hydrometer to evaluate the condition of the battery. The observation that a substance sinks or floats in a liquid is determined by the densities of the two substances.

EXPERIMENT 5

To determine the density of a substance, both its mass and the volume must be measured. You have carried out both of these procedures in previous experiments. From the mass and volume, the density is calculated. If the mass is measured in grams, and the volume in milliliters, the density will have the units of g/mL.

$$\text{Density of a substance} = \frac{\text{mass of substance}}{\text{volume of substance}} = \frac{\text{g substance}}{\text{mL substance}}$$

Specific Gravity

The specific gravity of a substance is a comparison of the density of that substance with the density of water which is 1.00 g/ml (4°C).

$$\text{specific gravity (sp gr)} = \frac{\text{density of substance (g/mL)}}{\text{density of water (g/mL)}}$$

Specific gravity is a number with no units; the units of density (g/mL) have canceled out. This is one of the few measurements in chemistry written without any units.

LABORATORY ACTIVITIES

A. DENSITY OF A SOLID

BE SURE YOU PUT ON YOUR SAFETY GOGGLES BEFORE YOU BEGIN THIS EXPERIMENT!

A.1 **Mass of a Solid.** Obtain two solid objects of different materials. Weigh each on the laboratory balance. Record the mass of each. (Remember that your measurements must have two figures after the decimal points as well as units.)

A.2 **Volume of a Solid** To determine the volume of each solid, volume displacement will be used. Fill a graduated cylinder about half full of water and record the water level. Carefully place one of the solid objects in the water. Record the level of the water after the solid is completely submerged. Calculate the volume of each by subtracting the initial water level from the final water level.

A.3 **Calculating the density** Show your calculations for density (g/mL) of each solid by dividing the mass (g) by the volume (mL).

$$\text{Density of solid} = \frac{\text{mass (g) of solid}}{\text{volume (mL) of solid}}$$

DENSITY AND SPECIFIC GRAVITY

B. DENSITY OF A LIQUID

B.1 **Volume of a Liquid.** Place some water or other liquid in a graduated cylinder. Record the volume of the liquid.

B.2 **Mass of a Liquid** Weigh a clean, dry, small beaker and record its mass. Transfer the liquid in the cylinder to the beaker and weigh again. Calculate the mass of liquid by difference.

B.3 **Density of a Liquid** Calculate the density of the liquid by dividing the mass (g) of the liquid by the volume (mL) of the liquid that was used. Repeat steps B.1-B.3 with a second liquid.

$$\text{Density of liquid} = \frac{\text{mass (g) of liquid}}{\text{volume (mL) of liquid}}$$

C. SPECIFIC GRAVITY

C.1 To calculate the specific gravity (sp gr) of the liquids in B, divide each density by the density of water which is 1.00 g/mL at 4°C.

$$\text{specific gravity of liquid} = \frac{\text{Density of liquid (g/mL)}}{\text{Density of water (1.00 g/mL)}}$$

Using a Hydrometer Specific gravity can be measured with a hydrometer. Small hydrometers (urinometers) are used in the hospital to determine the specific gravity of urine. Another type of hydrometer is used to measure the specific gravity of the fluid in your car battery. A hydrometer placed in a liquid is spun slowly to keep it from sticking to the sides of the container. The specific gravity scale is read at the meniscus of the fluid. Read the hydrometer scale to the one-thousandths place, three figures after the decimal point. (See Figure 5.1.) Some hydrometers use the European decimal point which is a comma. Record this as a decimal point.

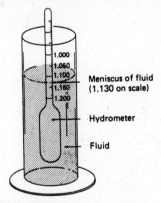

FIGURE 5.1 Reading a hydrometer.

C.2 Read the hydrometers set in graduated cylinders containing the same liquids you used in the density section. Record the hydrometer readings for water and other liquids.

C.3 Compare the calculated specific gravities (C.1) with the hydrometer readings for specific gravity.

EXPERIMENT 5

D. GRAPHING MASS AND VOLUME OF A LIQUID

In this experiment, we will illustrate the relationship between mass and volume in a graph. Four or five different samples of the same substance will be used. The volume, mass and density will be determined for each. After the data for several samples is collected, the mass and volume of each sample will be used to prepare a graph. See Construction of a Graph in the Appendix.

D.1 Weigh a dry, clean beaker and record its mass.

D.2 Obtain 5.0 mL of a liquid. You may use a 5.0 or 10.0-mL graduated cylinder or a 5.0-mL pipet. Transfer the 5.0 mL of liquid to the beaker. Weigh the beaker and the liquid it contains. Record the volume and the mass of the beaker and liquid.

For the second sample, measure out another 5.0 mL of liquid and add to the previous amount of liquid in the beaker. This give a new volume of 10.0 mL. Record the new mass. Repeat again adding another 5.0 mL of liquid to the beaker and weighing the beaker and liquid again. Continue until you have added a total of 25.0 mL of liquid to the beaker.

D.3 Calculate the mass of each sample of liquid by subtracting the mass of the beaker.

D.4 For each sample, calculate the density of the liquid.

D.5 On the graph, plot the data for each mass and volume measurement. The spaces on each axis are divided into equal intervals that go from zero to the highest volume or mass. Draw a smooth line through the points.

NAME_____ SECTION_____ DATE_____

EXPERIMENT 5
MEASURING DENSITY AND SPECIFIC GRAVITY
LABORATORY REPORT

A. DENSITY OF A SOLID

A.1 Object   _____   _____

     Mass   _____g   _____g

A.2 Volume of Water + object   _____mL   _____mL

     Volume of water   _____mL   _____mL

     Volume of object   _____mL   _____mL

A.3 Density of solid   _____g/mL _____g/mL

     Calculations for density:

B.   DENSITY OF A LIQUID

B.1 Type of liquid   _____   _____

     Volume of liquid   _____mL   _____mL

B.2 Mass of beaker + liquid   _____g   _____g

     Mass of beaker   _____g   _____g

     Mass of liquid   _____g   _____g

B.3 Density of liquid   _____g/mL _____g/mL

     Calculations for density:

EXPERIMENT 5

C. SPECIFIC GRAVITY

C.1 Type of liquid _____   _____

Density from B.3 _____ g/mL   _____ g/mL

Specific gravity
(calculated) _____   _____

Calculations for specific gravity:

C.2 Type of liquid _____   _____

Specific gravity
(hydrometer reading) _____   _____

How does the specific gravity calculated from density compare to the specific gravity from reading a hydrometer?

D. GRAPHING MASS AND VOLUME OF A LIQUID

D.1 Type of liquid _____

Mass of beaker _____ g

D.2 Mass of beaker and liquid	D.3 Mass of liquid	Volume	D.4 Density
(1) _____ g	_____ g	5.0 mL	_____ g/mL
(2) _____ g	_____ g	10.0 mL	_____ g/mL
(3) _____ g	_____ g	15.0 mL	_____ g/mL
(4) _____ g	_____ g	20.0 mL	_____ g/mL
(5) _____ g	_____ g	25.0 mL	_____ g/mL

D.5 Graph: Relationship of Mass and Volume

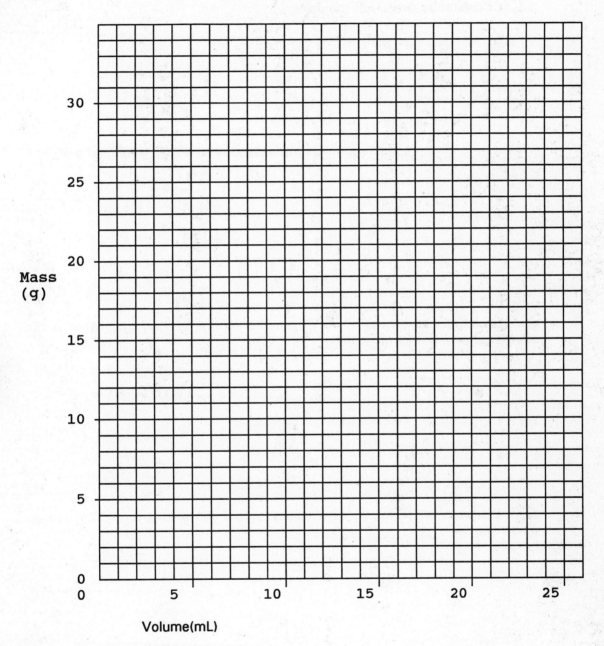

What is the relationship shown between mass and volume in the above graph?

EXPERIMENT 5

QUESTIONS AND PROBLEMS

1. An object made of aluminum has a mass of 8.37 g. When it was placed in a graduated cylinder containing 20.0 mL of water, the water level rose to 23.1 mL. Calculate the density of the object.

_____ g

2. What is the mass of a solution that has a density of 0.775 g/mL and a volume of 50.0 mL?

_____

3. A salt solution has a mass of 80.25 g and a volume of 75.0 mL. What is the specific gravity of that solution?

_____

4. From the graph in Section D, what volume would you predict for a 12 g sample of the liquid?

5. From the graph in Section D, what mass would you predict has a volume of 18 mL?

EXPERIMENT 6
TEMPERATURE AND CHANGES OF STATE

GOALS

1. Operate a Bunsen burner properly.
2. Measure temperature using the Celsius temperature scale.
3. Convert a temperature reading to its Fahrenheit and Kelvin temperatures.
4. Graph a change in temperature with time.
5. Identify change (s) of state on a heating and a cooling curve.

MATERIALS NEEDED

Bunsen burner
thermometer
matches or a striker
ice
250 or 400 mL beaker
rock salt

wire gauze
ringstand and iron ring
buret clamp
timer (or watch with second hand)
stirring apparatus setup (large test tube containing phenylsalicylate (salol, two-hole stopper with a thermometer and wire stirrer)

CONCEPTS TO REVIEW

Celsius temperature scale
Kelvin temperature scale
temperature conversion equations
graphing
changes of state
heating and cooling curves

BACKGROUND DISCUSSION

Temperature is a measure of the intensity of heat in a substance. A substance with le heat feels cold. Where the heat intensity is great, a substance feels hot. The temperature of our bodies is an indication of the heat produced. An infection may cause body temperature to deviate from normal. In the laboratory, temperature is measured on the Celsius scale. Celsius temperature can be converted to the corresponding Fahrenheit temperature using the following equation:

$$°F = 1.8 (°C) + 32$$

A Celsius temperature can be converted to Kelvin temperature by using the equation:

$$K = °C + 273$$

The particles of a liquid are in constant motion. When the liquid is heated, some molecules gain enough energy to escape from the liquid. When the molecules begin to form bubbles within the liquid, we say that the liquid is **boiling**. If the liquid is cooled, the particles move more slowly, eventually arranging themselves into a regular pattern characteristic of the solid structure of that substance. This change of state is called *freezing*.

A change of state becomes more obvious when the temperature of the substance is plotted against time. When a liquid boils, the temperature remains constant and a horizontal line or plateau appears on the graph. The temperature at which that plateau occurs is the **boiling point** of the liquid. It will not increase as long as the liquid continues to boil. A plateau also appears on a cooling curve of a liquid indicating the change of state from liquid to solid which occurs at the *freezing point*.

LABORATORY ACTIVITIES

A. THE LABORATORY BURNER

In most chemistry laboratories, heat is supplied by a Bunsen burner. While there are several models, they are similar in design. A burner is connected to a natural gas source at the lab bench by a piece of rubber tubing. The amount of gas going to the burner can be regulated by opening or closing the gas lever at the bench. The flow of gas can also be controlled by a wheel or screw at the base of the burner. Turning this wheel or screw into the burner decreases the flow of gas; turning it out increases the flow of gas. See Figure 6.1.

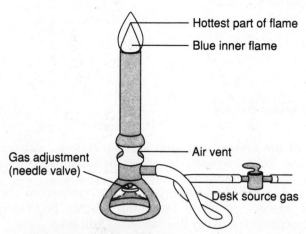

FIGURE 6.1 *A typical laboratory burner.*

When the gas enters the lower section of the burner, it passes through the air intake vent, and into the barrel of the burner. The amount of air that enters the mixture is adjusted by turning the barrel which opens or closes the air intake vents. The gas

TEMPERATURE AND CHANGE OF STATE

mixture is ignited at the top of the burner as it leaves the barrel. Make sure the gas valves are tightly closed when you leave the laboratory after using the Bunsen burner. See Figure 6.1.

> **SAFETY NOTE: A BUNSEN BURNER IS A POTENTIAL HAZARD!** Be sure that long hair is tied back. Keep the work area clear of books, papers, backpacks, and other flammable items. Always place the burner on the desk top. Never stack books up to set the burner at a higher level.

BE SURE YOU HAVE YOUR SAFETY GOGGLES ON BEFORE YOU BEGIN THIS EXPERIMENT!

A.1 Before you light the burner, practice the following:
 (a) opening and closing the *gas lever at the lab bench*
 (b) opening and closing the *gas valve at the base of the burner*
 (c) opening and closing the *air intake vents*

Close the holes of the air intake vents. Prepare to ignite the burner by having a striker or match ready. Your instructor may demonstrate the use of the striker. Turn on the gas and hold the flame or spark at the top of the burner. If you are using a striker, be sure you strike the flint hard enough to get sparks. Practice lighting the burner a few times.

A.2 When the air intake vents are closed, the gas mixture does not have an adequate supply of oxygen. The orange, sooty flame indicates the incomplete combustion of the gas. Open the air vents and increase the amount of air that mixes with the gas. The color of the flame will change to blue. Continue adjusting the air supply until the flame consists of two parts, an inner cone and an outer flame.

A.3 For general heating, you obtain the most heat from the tip of the inner blue flame. This is the part of the flame that should be in contact with any items you heat. Opening the air vents too far will make the burner noisy or will blow out the flame. Answer the questions in the laboratory record.

B. MEASURING TEMPERATURE

The laboratory thermometer is larger than the thermometer you use at home or in the hospital. The liquid in a laboratory thermometer responds quickly to the surroundings. When you are measuring the temperature of a solution or substance, always read the thermometer while it is **immersed** in that solution or substance.

> **SAFETY NOTE: NEVER SHAKE DOWN A LABORATORY THERMOMETER!** There is no need to shake a laboratory thermometer because it contains no constrictions as does an oral thermometer. Shaking a laboratory thermometer can cause breakage and serious accidents.

B.1 Observe the markings on a thermometer. Answer the questions on the record sheet.

B.2 To measure the temperature of a liquid, place the bulb of the thermometer in the center of the solution. When the temperature becomes constant, record the temperature (°C). Prepare the following and measure the temperature of each:

EXPERIMENT 6

  a. *Room temperature* obtained by placing the thermometer on the lab bench.
  b. Enough *tap water* to fill a beaker 1/3 full.
  c. An *ice slurry* prepared by adding ice to the water in **part b** to double (approx.) the volume.
  d. A *salted ice slurry* prepared by adding rock salt to the ice slurry in **part c**.

Complete the temperature table by converting the Celsius temperatures to their corresponding temperatures on the Fahrenheit and Kelvin scales.

C. OBSERVING AND GRAPHING A HEATING CURVE

Collecting Temperature Measurements   Set up a hot water bath as shown in Figure 6.2 using an iron ring, a wire gauze, a 250 or 400 mL beaker, a thermometer, and a buret clamp. The height of the iron ring should be about 1½-2 inches above the burner. (Recall that the tip of the blue inner flame should touch the bottom of the beaker for efficient heating.) Do not let the thermometer rest on the side or bottom of the beaker. It can be supported by tying a string to the loop in the top of the thermometer or placing the thermometer securely in a buret clamp. Adjust the thermometer so that the bulb is in the center portion of the ice or liquid.

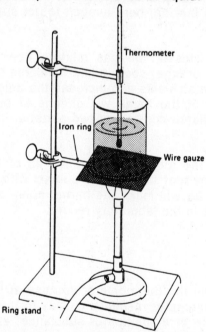

FIGURE 6.2   *Setup of apparatus for hot water bath.*

C.1   Fill the beaker half to two-thirds full with approximately equal parts crushed ice (or ice cubes) and water. Record the temperature of the ice-water slurry when the temperature becomes constant. This first measurement is recorded at a time of 0 min. Begin heating the water. Using a watch with a second hand or a timer obtained from the stockroom, record the temperature of the water every minute. Eventually the water will come to a *full boil*. (The appearance of small bubbles of escaping gas does not indicate boiling.) Continue to record the temperature of boiling water for another five minutes. (The temperature of boiling water should be almost constant for the remaining readings.) If necessary, add more time to the data table in C.1.

C.2 **Graphing the Heating Curve** Using the data you collected in C.1, construct a heating curve for water. See Figure 6.3. Label the areas of solid, liquid and boiling on the graph. The plateau for constant temperature during boiling indicates the *boiling point* of the water.

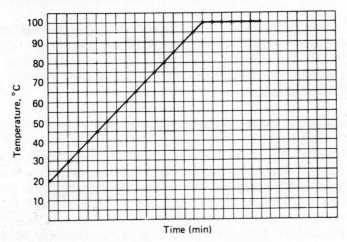

FIGURE 6.3 *Example of a graph for plotting a heating curve.*

D. OBSERVING AND GRAPHING A COOLING CURVE

The substance called phenylsalicylate (salol) is solid at room temperature. Since this substance is difficult to clean out of the test tube, your instructor may have test tubes with salol and stirring set ups *already prepared*. See Figure 6.4. If so, obtain one of these set ups from your instructor. **DO NOT TRY TO MOVE THE THERMOMETER WHILE IT IS FROZEN IN THE SOLID SUBSTANCE; IT WILL BREAK.**

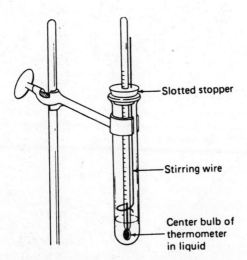

FIGURE 6.4 *Stirring apparatus for freezing point determination.*

If a stirring setup is not available, prepare one by placing phenylsalicylate in the bottom of a dry test tube to a depth of about 8 cm. The stirring apparatus must fit the test tube. The stirring apparatus consists of a double-holed slotted stopper with a thermometer carefully placed in the slotted hole. The stirring wire goes through the other hole. The bulb of the thermometer must be in the center of the liquid after the phenylsalicylate melts. The loop of the stirring wire goes around the thermometer.

BE EXTREMELY CAREFUL WHEN YOU PLACE THE THERMOMETER IN THE STOPPER. WRAP A PAPER TOWEL AROUND THE LOWER PORTION OF THE THERMOMETER. HOLD THE THERMOMETER CLOSE TO THE BULB, WET THE STOPPER, AND VERY SLOWLY BEGIN TO INSERT THE THERMOMETER.

D.1   Place the test tube setup containing salol attached to a buret clamp in the beaker of hot water from part C. (If it does not entirely melt, you may need to heat the water with the Bunsen burner.) When all the salol in the test tube has melted and the temperature of the liquid reaches 65°-75°C, remove the stirring apparatus from the hot water. Holding the test tube and contents in the air, record the temperature of the salol at 0 minutes. Use the wire stirrer to mix the contents and record the temperature every minute on the data table in D.1. As the salol begins to freeze (solidify), stop stirring. Continue to take temperature readings for another 10 minutes.

You may notice a jump in temperature just before the salol freezes. This condition called *supercooling* occurs when the temperature of the liquid goes below its freezing point without freezing. See Figure 6.5. When the particles of the liquid substance are arranged properly to form the solid structure, the temperature reverts to its freezing point. A constant temperature for several readings indicates the *freezing point* of the salol. When all of the substance has frozen, the temperature will drop again. However, it cannot cool any further than room temperature. Return the stirring apparatus and contents to the instructor when you are finished with this experiment.

D.2   Plot the cooling curve for salol on the graph provided in the laboratory record. Indicate the areas of liquid and solid, super-cooling (if any), and the freezing point for the substance. An example of a cooling curve is given in Figure 6.5.

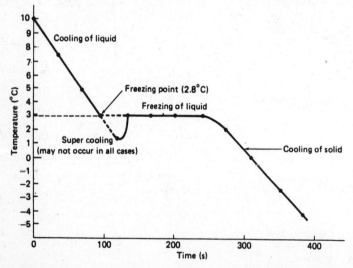

FIGURE 6.5   *A typical cooling curve showing supercooling.*

NAME_____ SECTION_____ DATE_____

EXPERIMENT 6
TEMPERATURE AND CHANGES OF STATE
LABORATORY REPORT

A. THE LABORATORY BURNER

A.1   How do you control the height of the flame?

A.2   What happens to the color of the flame as the air vent is opened?

A.3   What part of the flame should you use for general heating?

B. MEASURING TEMPERATURE

B.1   What temperature scale (s) is/are represented on the thermometer?

What are the highest and the lowest temperatures that can be measured using your laboratory thermometer?

EXPERIMENT 6

B.2 Substance   TEMPERATURE
                    °C   °F   K

a. Room temperature _____   _____   _____

b. Tap water   _____   _____   _____

c. Ice slurry   _____   _____   _____

d. Salted ice slurry _____   _____   _____

C. OBSERVING AND GRAPHING A CHANGE IN TEMPERATURE

C.1 *Heating Curve Data Table*

Time(min)	Temperature (°C)	Time(min)	Temperature (°C)
0	_____	_____	_____
1	_____	_____	_____
2	_____	_____	_____
_____	_____	_____	_____
_____	_____	_____	_____
_____	_____	_____	_____
_____	_____	_____	_____
_____	_____	_____	_____
_____	_____	_____	_____
_____	_____	_____	_____
_____	_____	_____	_____
_____	_____	_____	_____
_____	_____	_____	_____
_____	_____	_____	_____

NAME_____ SECTION_____ DATE_____

C.2 Graphing the Heating Curve for Water

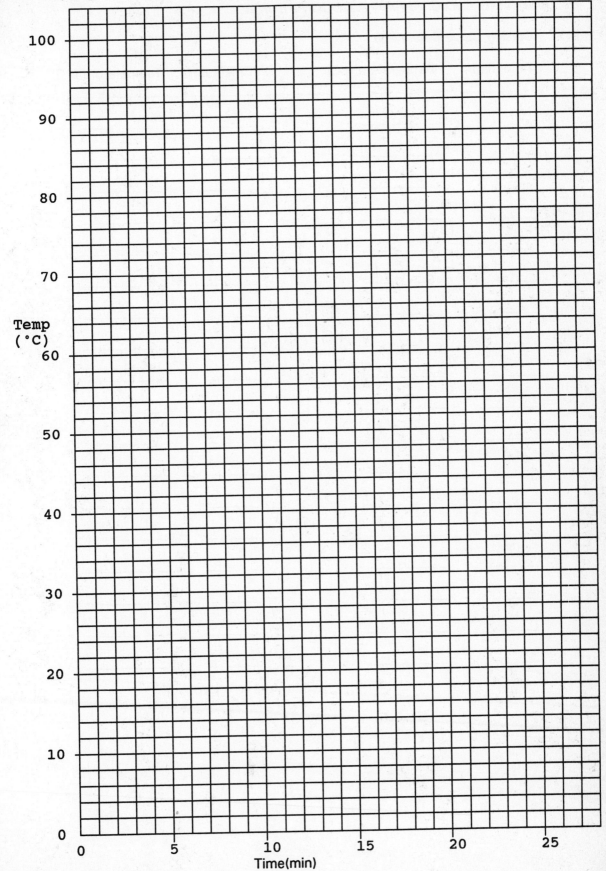

EXPERIMENT 6

D. OBSERVING AND GRAPHING A COOLING CURVE

D.1 *Cooling Curve Data Table*

Time(min)	Temperature (°C)	Time(min)	Temperature (°C)
0	_____	_____	_____
1	_____	_____	_____
2	_____	_____	_____
_____	_____	_____	_____
_____	_____	_____	_____
_____	_____	_____	_____
_____	_____	_____	_____
_____	_____	_____	_____
_____	_____	_____	_____
_____	_____	_____	_____
_____	_____	_____	_____
_____	_____	_____	_____
_____	_____	_____	_____
_____	_____	_____	_____
_____	_____	_____	_____
_____	_____	_____	_____
_____	_____	_____	_____
_____	_____	_____	_____

NAME_____ SECTION_____ DATE_____

D.2 *Graphing the Cooling Curve for Salol*

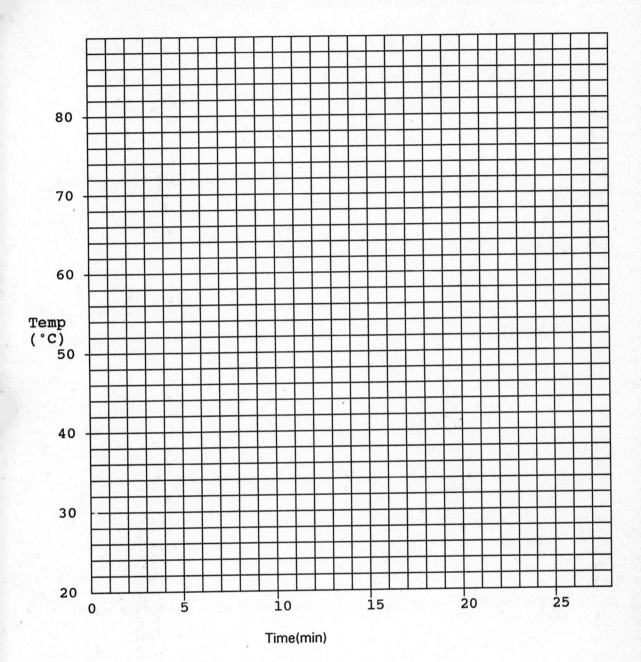

On the cooling curve, what is the freezing point of salol?

EXPERIMENT 6

QUESTIONS AND PROBLEMS

1. Write an equation for each of the following temperature conversions:

    a. °C to °F

    b. °F to °C

    c. °C to K

2. A recipe calls for a baking temperature of 205°C. What °F temperature should be set?

    _____°F

3. A laboratory experiment is run at —125°F. What is that temperature in kelvins?

    _____K

4. How would a burner with a larger flame affect the time required to reach the boiling point of water?

5. On the heating curve for water, how many minutes did it take for the temperature to rise to 60°C?

6. Eventually, during heating, the temperature reaches a constant value and forms a plateau on the graph. What does the plateau indicate?

EXPERIMENT 7
MEASURING HEAT OF REACTION

GOALS

1. Observe a temperature change during a chemical reaction.
2. Use a temperature change to calculate the heat of reaction.
3. Identify a reaction as endothermic or exothermic.
4. Use calorimetry to determine the caloric value of a food.

MATERIALS NEEDED

styrofoam cup
cardboard cover
thermometer(°C)
$NH_4NO_3(s)$, $CaCl_2(s)$ anhydrous
aluminum can
wire gauze
food such as cheese puffs, potato chips, or corn chips

CONCEPTS TO REVIEW

calorie
specific heat of water
calculation of heat energy
caloric values

BACKGROUND DISCUSSION

Chemical reactions typically involve a loss or gain of heat. If heat is released by the reaction, we say that the reaction is **exothermic**. If heat is absorbed by the reaction, we say that the reaction is **endothermic**.

In this experiment, we will determine the **heat of solution** for the dissolving of salts in water. The change in temperature during the solution process is utilized in first aid packs called hot packs and cold packs. Each type of pack contains water, and a salt in a vial. When the pack is hit or squeezed, the vial breaks, and the substance dissolves in the water causing a temperature change. If the change is exothermic, the pack becomes hot; if the change is endothermic, the pack becomes cold.

exothermic: salt(s) ⟶ salt(aq) + heat of solution
endothermic: salt(s) + heat of solution ⟶ salt(aq)

EXPERIMENT 7

Two salts used in hot packs and cold packs are ammonium nitrate, NH_4NO_3, and calcium chloride, $CaCl_2$. In this experiment, you will determine which substance is used in a hot pack and which is used in a cold pack.

The heat involved in a reaction is measured by **calorimetry**. In this experiment, the **calorimeter** will be a styrofoam cup. See Figure 7.1.

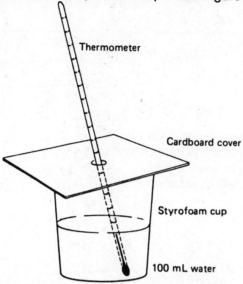

FIGURE 7.1 *Styrofoam cup setup for calorimetry.*

By measuring the mass of the salt, the mass of water in the calorimeter, and the temperature change of the water, the heat of solution in calories/gram can be calculated. For this we use the **specific heat** of water which is 1.0 **calorie/g°C**, the amount of heat required to raise the temperature of 1 gram of water by 1°C.

specific heat of water = $\dfrac{1.0 \text{ calorie}}{(1 \text{ g})(1°C)}$

heat (calories) = **g water** x **temperature change** x **specific heat**

For example, we can calculate the heat in calories required to raise the temperature of 20.0 g of water from 18.0°C to 24.0°C.

temperature change = 24.0°C - 18.0°C = 6.0°C

heat (calories) = 20.0 g x 6.0°C x $\dfrac{(1.0 \text{ calorie})}{(1 \text{ g})(1°C)}$ = 120 cal

When we know the quantity of heat released or absorbed by a salt, we can determine the heat of solution in calories per gram of salt. For example, suppose that the heat calculated above was the result of dissolving 2.50 g of a salt in water. Its heat of solution would be calculated as follows:

heat of solution = $\dfrac{120 \text{ cal}}{2.50 \text{ g}}$ = 48 cal/g

Since the temperature of the water increased, we would say that this reaction was exothermic.

In the second part of the experiment, we will see how the calories from foods are measured. Burning a cheese puff or some chips releases heat that can be used to heat water in an aluminum can. In this experiment, some of the heat is lost to the can and the surrounding air, but you can still get an idea of how nutritionists determine the caloric content of foods. Food calories(Calories) are actually kilocalories; caloric values are reported as kcal/g of food.

1 kilocalorie = 1000 calories

Type of food	Caloric Value
carbohydrate	4 kcal/g
fat	9 kcal/g
protein	4 kcal/g

LABORATORY ACTIVITIES

 WEAR PROTECTIVE GLASSES!

A. HEAT OF SOLUTION

A.1 Weigh a calorimetry cup. Record its mass to the nearest 0.1 g. Add about 50 mL of water to the cup and reweigh.

A.2 Weigh a watch glass. Record its mass to the nearest 0.1 g. Add approximately 4 g of ammonium nitrate(NH_4NO_3) crystals. Reweigh.

A.3 Record the temperature of the water in the calorimetry cup. Add the salt and replace the cover. Stir gently to dissolve all of the salt. Record the lowest or highest temperature reached by the solution.

Empty out the contents of the styrofoam cup, and rinse. **Repeat steps A.1-A.3 using anhydrous calcium chloride (s).**

Calculations:

A.4 Calculate the mass of water in the calorimetry cup. Subtract the mass of the cup from the combined mass of the cup and water.

A.5 Calculate the mass of the salt.

A.6 Calculate the *change* in temperature(°C) for each sample.

A.7 Calculate the calories for the heat of solution.

$$\text{heat(calories)} = \text{g water} \times \text{temperature change} \times \frac{1 \text{ cal}}{(1 \text{ g})(1 °C)}$$

EXPERIMENT 7

A.8 Calculate the heat of solution. Divide the calories(A.7) by the mass of the salt(A.5).

$$\text{heat of solution} = \frac{\text{number of calories}}{\text{number of grams}} = \text{cal/g}$$

A.9 Describe the reaction as exothermic or endothermic. State whether the salt would be used in a hot pack or a cold pack.

A.10 The actual values for the salts are given below.

NH_4NO_3 77 cal/g
$CaCl_2$ 162 cal/g

Calculate **percent error** by obtaining the difference (without a sign) between the actual value and your experimental value, dividing by the actual value, and multiplying by 100.

$$\text{percent error} = \frac{|\text{actual} - \text{experimental}|}{\text{actual}} \times 100$$

B. MEASURING THE CALORIC VALUE OF A FOOD

B.1 Obtain an aluminum can and weigh it to the nearest 0.1 g. Add about 100 mL of water and reweigh. Place the can in a setup as shown in Figure 7.2. Place the aluminum can on an iron ring. Place a second iron ring with a wire gauze a short distance below the aluminum can. Suspend a thermometer from a clamp so the bulb is below the water level in the aluminum can.

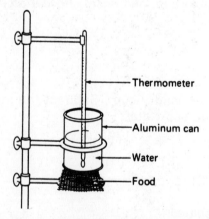

FIGURE 7.2 Setup for determination of the caloric value of a food.

B.2 Weigh the food sample you are going to use and record to the nearest 0.01 g.

B.3 Record the temperature of the water in the can. Place the food sample on the wire gauze and ignite it. Remove the heat source immediately and let the food sample burn. Record the highest temperature reached by the water in the can.

B.4 Weigh any remaining food sample to 0.01 g and record.

Calculations

B.5 Calculate the mass of water in the aluminum can by subtracting the mass of the can from the combined mass.

B.6 Calculate the temperature change of the water after it has been heated by the combustion reaction.

B.7 Calculate the calories.

$$\text{calories} = \text{mass of water} \times \text{temperature change} \times \frac{1.0 \text{ calorie}}{(1 \text{ g})(1 \,^\circ\text{C})}$$

B.8 Convert the calories from B.7 to kilocalories.

$$\text{kcal} = \text{number of cal} \times \frac{1 \text{ kcal}}{1000 \text{ cal}}$$

B.9 Calculate the mass of food sample that reacted. Subtract the mass of any remaining food after combustion from the mass of the food sample.

B.10 Determine the caloric value of the food by dividing the kilocalories by the grams of the food sample that reacted (B.8 ÷ B.9).

$$\text{caloric value} = \frac{\text{kcal}}{\text{g food}}$$

NAME_____ SECTION_____ DATE_____

EXPERIMENT 7
MEASURING HEAT OF REACTION
LABORATORY REPORT

A. HEAT OF SOLUTION

		NH_4NO_3	$CaCl_2$
A.1	Calorimeter + water	_____ g	_____ g
	Calorimeter (cup)	_____ g	_____ g
A.2	Watch glass + salt	_____ g	_____ g
	Watch glass	_____ g	_____ g
A.3	Final water temperature	_____ °C	_____ °C
	Initial water temperature	_____ °C	_____ °C

Calculations:

		NH_4NO_3	$CaCl_2$
A.4	Mass of water	_____ g	_____ g
A.5	Mass of salt	_____ g	_____ g
A.6	Temperature change	_____ °C	_____ °C
A.7	Heat in calories *Show calculations*	_____ cal	_____ cal
A.8	Heat of solution *Show calculations*	_____ cal/g	_____ cal/g
A.9	Type of reaction	_____	_____
	Hot pack or cold pack	_____	_____
A.10	Percent error *Show calculations*	_____ %	_____ %

EXPERIMENT 7

B. MEASURING THE CALORIC VALUE OF A FOOD

Type of Food sample _____

B.1 Aluminum can + water _____ g

Aluminum can _____ g

B.2 Mass of food _____ g

B.3 Final temperature of water _____ °C

Initial temperature of water _____ °C

B.4 Mass of food remaining _____ g

Calculations:

B.5 Mass of water _____ g

B.6 Temperature change _____ °C

B.7 Heat in calories _____ cal
Show calculations

B.8 Heat in kilocalories _____ kcal

B.9 Mass of food reacted _____ g

B.10 Caloric value _____ kcal/g
Show calculations

NAME_____ SECTION_____ DATE_____

QUESTIONS AND PROBLEMS

1. Write the standard expression and value for the specific heat of water(liquid).

2. Water has one of the largest specific heats of any substance. Why is this important for the human body?

3. Why do materials such as copper, silver or tin tend to get hot quickly when heated?

4. How many calories are required to raise the temperature of 225 g of water from 42°C to 75°C?

5. A 0.1 g sample of a pretzel is burned. The heat it gives off is used to heat 50.0 g of water from 25.0°C to 30.0°C. What is the caloric value of the pretzel in kcal/g?

EXPERIMENT 8
MEASURING HEAT OF FUSION AND VAPORIZATION

GOALS

1. Measure the temperature change when ice is added to water.
2. Calculate the heat of fusion for water.
3. Measure the temperature change when steam is added to water.
4. Calculate the heat of vaporization for water.

MATERIALS NEEDED

calorimeter (styrofoam cup and cover)
thermometer
ice cubes

Demonstration by Instructor:
stopper (1-hole) fitted with glass tubing to fit an Erlenmeyer flask
buret clamp
rubber tubing attached to glass tubing (15 cm)
calorimeter
thermometer

CONCEPTS TO REVIEW

calories
calorimetry
calculation of heat energy
heat of fusion
heat of vaporization

EXPERIMENT 8

BACKGROUND DISCUSSION

The amount of heat required to melt (or freeze) a substance is called the *heat of fusion*.

$$H_2O(s) + \text{heat} \longrightarrow H_2O(l)$$

The standard value for the heat of fusion for water is 80 cal/g. As long as the solid is melting(or freezing), the temperature remains constant. In this experiment, a certain amount of water will be cooled by the melting of ice. By calculating the heat lost by the water, we can calculate the heat used to melt the ice. Ice can be added until the water temperature is close to 0°C. By lowering the water temperature to close to 0°C, the amount of heat used to raise the temperature of the melting solid will be negligible and need not be considered.

$$\text{heat of fusion} = \frac{\text{calories absorbed by ice}}{\text{grams of ice}}$$

When a substance changes from a gas to a liquid (or liquid to a gas) at the boiling point, the heat or energy required is called the *heat of vaporization*.

$$\text{heat} + H_2O(l) \longrightarrow H_2O(g)$$

When 1 g of steam changes to liquid water, 540 cal, the standard heat of vaporization, are released. Because this large quantity of heat could cause burns, this part of the experiment will be done as a demonstration by your instructor. Steam from a steam generator will be run through tubing into a measured quantity of water. The temperature increases as steam entering the water in the calorimeter changes to liquid. The difference in temperature will be used to calculate the amount of heat released.

$$\text{heat of vaporization} = \frac{\text{calories released by steam}}{\text{grams of steam}}$$

LABORATORY ACTIVITIES

Wear your safety goggles!

A. HEAT OF FUSION

A.1   Weigh the calorimetry cup and record its mass to the nearest 0.1 g. Add about 100 mL of water to the cup. Record the total mass in grams (0.1 g).

A.2   Measure the temperature of the water in the calorimeter. Record.

Place a medium-sized ice cube in the calorimeter. Stir gently. If the temperature does not drop to 3 or 4°C, add more ice. When a low temperature is reached, remove any unmelted ice cube. Record the final temperature.

A.3   Reweigh the calorimetry cup and water. The total mass will be increased by the ice that melted.

MEASURING HEAT OF FUSION AND VAPORIZATION

Calculations:

A.4 Calculate the mass of water added to the calorimeter.

A.5 Calculate the temperature change for the water.

A.6 Calculate the calories lost by the water. This value is equal to the heat absorbed by the ice.

$$\text{calories} = \text{mass of water} \times \text{temperature change} \times \frac{1.0 \text{ cal}}{g \, °C}$$

A.7 Calculate the mass of ice that melts. Subtract the mass of the cup and water from the mass of the cup, water, and water from the melted ice (A.3-A.1).

A.8 Calculate your experimental value for the heat of fusion for ice by dividing the calories by the mass of the ice that melted (A.6 ÷ A.7).

$$\text{heat of fusion} = \frac{\text{calories absorbed by ice}}{\text{grams of ice}}$$

A.9 Calculate the **percent error** by subtracting your experimental value from 80 cal/g, dividing by 80 cal/g, and multiplying by 100.

$$\text{percent error} = \frac{|80 \text{ cal/g} - \text{experimental cal/g}|}{80 \text{ cal/g}} \times 100$$

B. *HEAT OF VAPORIZATION* (Demonstration by Instructor)

A steam generator will be set up by your instructor that consists of a 250-mL Erlenmeyer flask fitted with a stopper (1-hole) and a short length of glass attached to a piece of tubing. Attached to the other end of the tubing is a piece of glass tubing about 15 cm long. The flask, about two-thirds full of water, is set on a wire screen and held in place by a buret clamp. When the water is boiling, steam will begin to flow out of the glass tubing. See Figure 8.1.

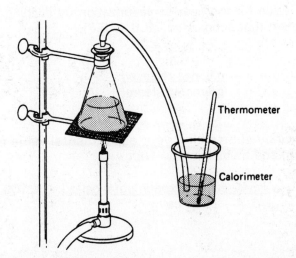

Figure 8.1. Apparatus for measuring the heat of vaporization.

EXPERIMENT 8

B.1 Weigh a calorimetry cup. Record its mass to the nearest 0.1 g. Place about 100 g of water in the cup and reweigh. Record the total mass.

B.2 Record the temperature of the water in the calorimeter.

CAREFUL! THIS MUST BE DONE BY YOUR INSTRUCTOR. HOT STEAM WILL BE COMING OUT OF THE GLASS TUBING. DO NOT TOUCH! STEAM HAS A HIGH HEAT OF VAPORIZATION AND CAN CAUSE BURNS.

When the water is boiling, the glass tubing will be placed in the water of the calorimeter. On the thermometer, follow the temperature as steam condenses inside the calorimeter. When the temperature reaches 75° or higher, remove the steam tubing, and turn off the burner under the boiling water.

Record the final temperature of the heated water in the calorimeter.

B.3 Reweigh the cup, water, and condensed steam. Record the total mass.

Calculations:

B.4 Calculate the mass of water added to the calorimeter.

B.5 Calculate the temperature change for the water.

B.6 Calculate the calories gained by the water. This value is equal to the heat released by the steam.

$$\text{calories} = \text{mass of water} \times \text{temperature change} \times \frac{1.0 \text{ cal}}{\text{g }°C}$$

B.7 Calculate the mass of steam that condensed. Subtract the mass of the calorimetry cup and water from the mass of the cup, water, and water from the condensed steam (B.3-B.1).

B.8 Calculate your experimental value for the heat of vaporization by dividing the calories by the mass of the steam that condensed (B.6 ÷ B.7).

$$\text{heat of vaporization} = \frac{\text{calories released}}{\text{grams of steam}}$$

A.9 Calculate the ***percent error*** by subtracting your experimental value from 540 cal/g, dividing by 540 cal/g, and multiplying by 100.

$$\text{percent error} = \frac{|540 \text{ cal/g} - \text{experimental cal/g}|}{540 \text{ cal/g}} \times 100$$

NAME_____ SECTION_____ DATE_____

EXPERIMENT 8
MEASURING HEAT OF FUSION
AND VAPORIZATION
LABORATORY REPORT

A. HEAT OF FUSION

A.1   Calorimeter + water   _____ g

        Calorimeter(cup)   _____ g

A.2   Final water temperature   _____ °C

        Initial water temperature   _____ °C

A.3   Calorimeter + water + melted ice   _____ g

Calculations:

A.4   Mass of water   _____ g

A.5   Temperature change   _____ °C

A.6   Heat in calories   _____ cal
        Show calculations

A.7   Mass of ice melted   _____ g

A.8   Heat of fusion   _____ cal/g
        Show calculations

A.9   Percent error   _____ %
        Show calculations

EXPERIMENT 8

B. HEAT OF VAPORIZATION

B.1 Calorimeter + water _____g

   Calorimeter(cup) _____g

B.2 Final water temperature _____°C

   Initial water temperature _____°C

B.3 Cup + water + condensed steam _____g

Calculations:

B.4 Mass of water _____g

B.5 Temperature change _____°C

B.6 Heat in calories _____cal
   Show calculations

B.7 Mass of steam added _____g

B.8 Heat of vaporization _____cal/g
   Show calculations

B.9 Percent error _____%
   Show calculations

NAME_____ SECTION_____ DATE_____

QUESTIONS AND PROBLEMS

1. How many calories are required to melt 25 g of ice at 0°C?

2. Draw a heating curve for water that begins with ice at 0°C and goes to liquid water at 75°C. Label each part.

3. Observe the heating curve in question 2 and calculate the kilocalories needed to convert 15 g of ice at 0°C to liquid water at 75°C.

4. Calculate the number of calories released when 50.0 g of steam condenses at 100°C.

5. Compare the amount of heat (kcal) released by 50.0 g of water at 100°C hitting the skin and cooling to body temperature (37°C) to 50.0 g of steam that hits the skin and cools to body temperature. Why are steam burns more severe?

EXPERIMENT 9
ATOMS AND ELEMENTS

GOALS

1. Write the correct symbols or names of some elements.
2. Describe some physical properties of the elements you observe.
3. Categorize an element as a metal or nonmetal by its physical properties or its location on the periodic table.
4. Given the isotope symbol of an atom, determine its mass number, atomic number, number of protons, neutrons, and electrons.
5. Write the isotope symbol from the mass number and atomic number of an atom.

MATERIALS NEEDED

a display of elements

CONCEPTS TO REVIEW

names and symbols of the elements
properties of metals and nonmetals
periodic chart
subatomic particles
isotopes

BACKGROUND DISCUSSION

Primary substances, called *elements*, build all the materials about you. There are 109 elements known today. Within every element, there are *atoms* of the element, the smallest units that are characteristic of that element. There are different atoms for each element.

In this experiment, you will be looking at some elements in a laboratory display. Some look very different from each other, while others look similar. Elements can be categorized in several ways. In this experiment, you are going to group elements by similarities in their physical properties. Elements that appear shiny or lustrous are called *metals*. Metals are usually good conductors of heat and electricity, somewhat soft and ductile, and can be molded into a shape. Some of the metals you will see such as

EXPERIMENT 9

sodium or calcium may have an outer coating of a white oxide formed by combination with oxygen in the air. If cut, you could see the fresh shiny metal underneath. Other elements called **nonmetals** are not good conductors of heat and electricity, are brittle, and appear dull, not shiny.

Atoms are made of smaller bits of matter called **subatomic particles**. Of these, we are interested in protons, neutrons, and electrons. **Protons** are positively charged particles, **electrons** are negatively charged, and **neutrons** are neutral (no charge). Within the atom, the protons and neutrons are tightly packed in the **nucleus**. The rest of the atom which is mostly empty space is occupied by fast-moving electrons. Electrons are so small that their mass is considered to be negligible compared to the mass of the proton or neutron.

Every atom can be identified by its atomic number and mass number. The number of protons is equal to the **atomic number** of the element.

$$\textit{atomic number} = \text{number of protons } (p^+)$$

Protons attract electrons because they have opposite charges. In a neutral atom, the number of protons is equal to the number of electrons.

$$\text{neutral atom}$$
$$\text{number of protons } (p^+) = \text{number of electrons } (e^-)$$

The **mass number** of an atom is the total number of protons and neutrons.

$$\textit{mass number} = \text{total number of protons and neutrons}$$

All atoms of the same element have the same number of protons, but they can differ in the number of neutrons. This means that atoms of the same element can have different atomic masses. Then the atoms of that element are called **isotopes**. The subatomic particles in an isotope are shown in the mass number and atomic number given in an **isotope symbol**.

Isotope symbol

mass number (p^+ and n^o) → 34
symbol of element → S
atomic number (p^+) → 16

LABORATORY ACTIVITIES

A. PHYSICAL PROPERTIES OF ELEMENTS

Complete the table in the laboratory record by writing the symbol and atomic number for each element. Observe the elements in the laboratory display of elements. Describe their physical properties (color and luster). From your observations, describe each element as a metal (M), or a nonmetal (NM).

ATOMS AND ELEMENTS

B. METALS AND NONMETALS ON THE PERIODIC TABLE

B.1   On the periodic table provided in the laboratory record, write the letter M on the symbols of those elements listed in Part A that are metals (shiny, ductile). Write NM on the symbols of the elements that are nonmetals (dull, not shiny, brittle, or are gases).

   Observe the location of the metals and the nonmetals relative to a heavy zig-zag line which looks like a staircase on the periodic chart. The terms metal and nonmetal are not really that definite, but the heavy zig-zag line helps to separate those elements with typically more metallic behavior from those elements that show more nonmetallic behavior. Use this information to answer the question in the laboratory record.

B.2   *Without looking at the display of elements*, use the location of the elements on the periodic table to *predict* whether the elements listed would be metals or nonmetals; shiny or dull. *After you have completed your predictions*, observe those same elements in the display to see if you predicted correctly.

C. SUBATOMIC PARTICLES AND ISOTOPES

C.1   Complete the table given in the laboratory record with the correct atomic numbers, mass numbers, number of protons, neutrons and electrons.

C.2   For each of the isotope symbols, state the number of protons, neutrons and electrons for each atom. Write the isotope symbol when given the number of subatomic particles.

NAME_____ SECTION_____ DATE_____

EXPERIMENT 9
ATOMS AND ELEMENTS
LABORATORY REPORT

A. PHYSICAL PROPERTIES OF ELEMENTS

Element	Symbol	Atomic number	Physical properties		
			color	luster	metal/nonmetal
zinc	_____	_____	_____	_____	_____
copper	_____	_____	_____	_____	_____
carbon	_____	_____	_____	_____	_____
oxygen	_____	_____	_____	_____	_____
aluminum	_____	_____	_____	_____	_____
nitrogen	_____	_____	_____	_____	_____
silver	_____	_____	_____	_____	_____
iron	_____	_____	_____	_____	_____
sulfur	_____	_____	_____	_____	_____
magnesium	_____	_____	_____	_____	_____
silicon	_____	_____	_____	_____	_____
tin	_____	_____	_____	_____	_____
_____	_____	_____	_____	_____	_____
_____	_____	_____	_____	_____	_____

EXPERIMENT 9

B. METALS AND NONMETALS ON THE PERIODIC TABLE

Group IA	2A												3A	4A	5A	6A	7A	8A
																		2 He 4.0
1 H 1.0																		
3 Li 6.9	4 Be 9.0												5 B 10.8	6 C 12.0	7 N 14.0	8 O 16.0	9 F 19.0	10 Ne 20.2
11 Na 23.0	12 Mg 24.3				Transition elements								13 Al 27.0	14 Si 28.1	15 P 31.0	16 S 32.1	17 Cl 35.5	18 Ar 39.9
19 K 39.1	20 Ca 40.1	21 Sc 45.0	22 Ti 47.9	23 V 50.9	24 Cr 52.0	25 Mn 54.9	26 Fe 55.8	27 Co 58.9	28 Ni 58.7	29 Cu 63.5	30 Zn 65.4		31 Ga 69.7	32 Ge 72.6	33 As 74.9	34 Se 79.0	35 Br 79.9	36 Kr 83.8
37 Rb 85.6	38 Sr 87.6	39 Y 88.9	40 Zr 91.2	41 Nb 92.9	42 Mo 95.9	43 Tc (99)	44 Ru 101.1	45 Rh 102.9	46 Pd 106.4	47 Ag 107.9	48 Cd 112.4		49 In 114.8	50 Sn 118.7	51 Sb 121.8	52 Te 127.6	53 I 126.9	54 Xe 131.3
55 Cs 132.9	56 Ba 137.3	57 La 138.9	72 Hf 178.5	73 Ta 180.9	74 W 183.8	75 Re 186.2	76 Os 190.2	77 Ir 192.2	78 Pt 195.1	79 Au 197.0	80 Hg 200.6		81 Tl 204.4	82 Pb 207.2	83 Bi 209.0	84 Po (210)	85 At (210)	86 Rn (222)

Note numbers in parentheses indicate mass number of most stable or best known isotope

B.1 Describe the location of the metals and nonmetals on the periodic table.

B.2 Element	Prediction Metal/nonmetal	Prediction Shiny/dull	Correct?
chromium (Cr)	_____	_____	_____
gold (Au)	_____	_____	_____
lead (Pb)	_____	_____	_____
boron (B)	_____	_____	_____
chlorine (Cl_2)	_____	_____	_____

NAME_____ SECTION_____ DATE_____

C. SUBATOMIC PARTICLES AND ISOTOPES

C.1 Element	Atomic number	Mass number	Neutrons	Protons	Electrons
fluorine	____	19	____	____	____
iron	____	____	30	____	____
____	____	27	____	____	13
____	____	____	20	19	____
bromine	____	80	____	____	____
____	____	197	____	____	79
____	____	____	74	53	____

C.2 Isotope symbol	Protons	Neutrons	Electrons
$^{40}_{20}Ca$	____	____	____
$^{43}_{20}Ca$	____	____	____
$^{44}_{20}Ca$	____	____	____
____	20	22	____
____	____	26	20

81

EXPERIMENT 9

QUESTIONS AND PROBLEMS

1. Complete the list of names of elements and symbols:

Element	Symbol	Element	Symbol
potassium	_____	_____	Na
sulfur	_____	_____	P
nitrogen	_____	_____	Fe
magnesium	_____	_____	Cl
copper	_____	_____	Ag

2. Compare the physical properties of metals and nonmetals.

3. Use the periodic table to categorize the following elements as metals or nonmetals.

   Na _____   S _____   Cu _____

   F _____   Fe _____   C _____

   Ca _____   O _____   Zn _____

4. a. An atom of chlorine has a mass number of 37. Describe how you can determine the number of neutrons in the chlorine atom.

   b. Another atom of chlorine has a mass number of 35. How does it compare to the above atom of chlorine?

5. A neutral atom has a mass number of 80 and 45 neutrons. Write its isotope symbol.

EXPERIMENT 10
PERIODIC PROPERTIES AND
ELECTRON ARRANGEMENT

GOALS

1. Draw a graph of atomic diameter against atomic number.
2. Interpret the trends in atomic radii within a family and a period.
3. Describe the color of a flame produced by an element.
4. Use the color of a flame to identify an element.
5. Draw a model of an atom including the electron arrangement for the first 20 elements.

MATERIALS NEEDED

For flame testing:
  platinum (or nichrome) wire
  Bunsen burner
  solutions (0.1 M)
    $CaCl_2$, KCl, $BaCl_2$, $SrCl_2$, $CuCl_2$, NaCl
  unknown solution (s)

6M HCl *(CAUTION: CORROSIVE)*
small beaker
spot plate

BACKGROUND DISCUSSION

Flame Tests

Certain elements produce strong colors in their flames when heated. Normally, there is a specific arrangement of the electrons in the energy levels occupied by the electrons. However, during heating one or more electrons may absorb energy in sufficient amounts (quanta) to jump to a higher (but less stable) energy level.

When an electron drops from a higher energy level to a lower energy level, energy is released. If the energy is emitted as visible light, the flame will be brightly colored. Each of these elements give a unique color in its flame, which can be used to identify an element.

The chemistry of an element strongly depends on the arrangement of the electrons. The energy levels for electrons of atoms of the first 20 elements have the following arrangement:

EXPERIMENT 10

Energy Level	Electrons
1	$2e^-$
2	$8e^-$
3	$8e^-$
4	$2e^-$

Periodic Properties: Variations in Atomic Radii

Since the 1800s, scientists have recognized that chemical and physical properties of certain groups of elements tend to be similar. A Russian scientist, Dmitri Mendeleev, found that the chemical properties of elements tended to occur periodically when the elements were arranged in order of increasing atomic mass. He used this periodic pattern to predict the behavior of elements that were not yet discovered. Later, H.G. Moseley established that the similarities in properties were associated with the atomic number.

In this exercise, you will be graphing the relationship between the atomic radius of an atom and its atomic number. Such a graph will show a repeating or periodic trend. In order to explain a graph of this type, observe the graph in Figure 10.1 that is obtained by plotting the average temperature of the seasons. A cycle of high and low temperatures that repeats each year. Such a tendency is known as a **periodic property**. There are three cycles on this particular graph, one full cycle occurring every year. When such cycles are known, the average temperatures for the next year could be predicted.

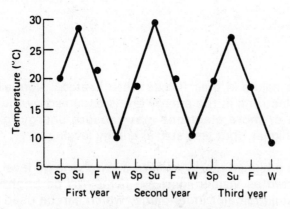

Figure 10.1 A graph of average seasonal temperatures.

PERIODIC PROPERTIES AND ELECTRON ARRANGEMENT

LABORATORY ACTIVITIES

A. FLAME TESTS

Be sure your safety goggles are on!

Preparation Obtain the materials for flame tests. Bend the end of the flame-test wire into a small loop and secure the other end in a small cork stopper. Pour a small amount of 6 M (6 N) HCl into a beaker and clean the wire by dipping the loop in the HCl.

6 M HCl is a corrosive acid. Be careful when you use it. Wash any spills on the skin with tap water for 10 minutes.

Adjust the flame of a Bunsen burner until it is colorless. Place the loop in the flame of the Bunsen burner. If you see a strong color in the flame while heating the wire, clean and heat the wire again until the color is gone.

Using the spot plate (a small plate with indentations), transfer small amounts of the solutions $CaCl_2$, KCl, $BaCl_2$, $SrCl_2$, $CuCl_2$, and NaCl in different spots. Be careful not to mix the different solutions. Label the solution on the spot plate diagram in the laboratory record.

Heating the flame test wire Dip the cleaned wire in the first solution on the spot plate. Make sure that a thin film of the solution adheres to the loop. (See Figure 10.2) Move the loop of the wire into the lower portion of the flame and record the color you observe. If the color is not intense enough, use a clean dropper to place a drop of the solution on the loop, and test again. The color of the KCl flame is short-lived. Be sure to observe the color of the flame from the KCl solution in the first few seconds of heating. Clean the wire and repeat the flame tests with the other solutions.

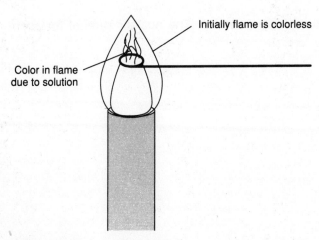

Figure 10.2 Flame test

EXPERIMENT 10

Testing an unknown solution Obtain an unknown solution that contains one of the above solutions. Using the flame test procedure, determine the color of the flame of your unknown. You may wish to compare the flame color to the color of the flame tests for the known solutions you looked at earlier. For example, if you think your unknown is KCl, recheck the color of the KCl solution for verification. Indicate the element that was responsible for the color of the flame for your unknown.

B. GRAPHING ATOMIC RADIUS VERSUS THE ATOMIC NUMBER

Atomic radii plotted against atomic numbers illustrate the concept of the periodic tendencies of the elements. The atomic radii for the first 25 elements are given in the laboratory record. Plot the atomic radius of each of the elements against the atomic number of the element on the graph.

Use the graph you prepare to answer questions in the laboratory record.

C. DRAWING MODELS OF ATOMS

A model of an atom can be drawn that shows the number of protons and neutrons in the nucleus with the electrons shown in energy shells. This is an oversimplification of the true nature of atoms, but the model does serve to illustrate the numerical relationship of the subatomic particles. For example, a drawing of an atom of boron with a mass number of 11 might look like this. Boron has atomic number 5 which means it has 5 protons. The number of neutrons is determined by subtracting the number of protons (5) from the mass number (11) to give 6 neutrons.

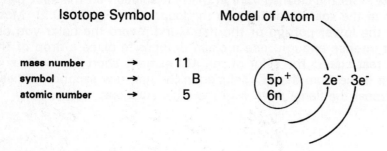

Draw a model of each atom listed on the laboratory record. The nuclear symbol for each atom is given.

NAME_____ SECTION_____ DATE_____

_____

EXPERIMENT 10
PERIODIC PROPERTIES AND
ELECTRON ARRANGEMENT
LABORATORY REPORT

A. FLAME TESTS

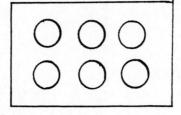

spot plate

Solution	Color of Flame
$CaCl_2$	_____
KCl	_____
$BaCl_2$	_____
$SrCl_2$	_____
$CuCl_2$	_____
NaCl	_____

Unknown Solution (s)	1	2	3
code number	_____	_____	_____
Color of flame	_____	_____	_____
Identification	_____	_____	_____

87

EXPERIMENT 10

B. GRAPHING ATOMIC RADIUS VERSUS ATOMIC NUMBER

Table 10.1 Atomic Radii for the Elements with Atomic Numbers 1-25

Element	Symbol	Atomic Number	Atomic Radius (pm) (picometer = 10^{-12}m)
first period			
hydrogen	H	1	37
helium	He	2	50
second period			
lithium	Li	3	152
beryllium	Be	4	111
boron	B	5	88
carbon	C	6	77
nitrogen	N	7	70
oxygen	O	8	66
fluorine	F	9	64
neon	Ne	10	70
third period			
sodium	Na	11	186
magnesium	Mg	12	160
aluminum	Al	13	143
silicon	Si	14	117
phosphorus	P	15	110
sulfur	S	16	104
chlorine	Cl	17	99
argon	Ar	18	94
fourth period			
potassium	K	19	231
calcium	Ca	20	197
scandium	Sc	21	160
titanium	Ti	22	150
vanadium	V	23	135
chromium	Cr	24	125
manganese	Mn	25	125

B. Graph of Atomic Radius versus Atomic Number

Atomic radius(pm)

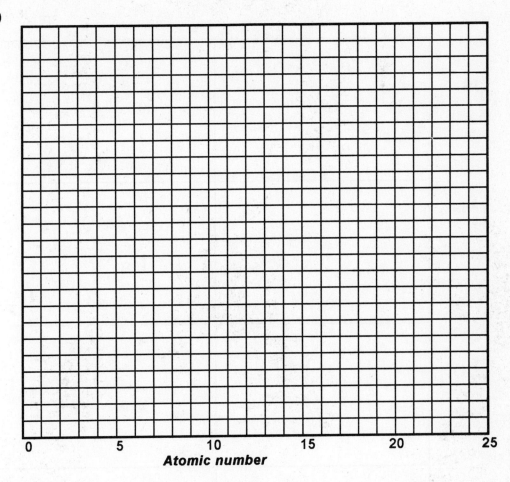

Atomic number

1. How do the atomic radii change from the left to the right in the second period from lithium (at.no. 3), to neon (at.no. 10)?

2. Would you expect a similar type of change in the third period from sodium to argon?

3. Why is the type of change seen in atomic radii called a periodic property?

EXPERIMENT 10

C. DRAWING MODELS OF ATOMS

$^{7}_{3}Li$			$^{14}_{7}N$
$^{20}_{10}Ne$			$^{19}_{9}F$
$^{25}_{12}Mg$			$^{1}_{1}H$
$^{24}_{11}Na$			$^{31}_{15}P$
$^{37}_{17}Cl$			$^{35}_{17}Cl$

NAME_____ SECTION_____ DATE_____

QUESTIONS AND PROBLEMS

1. You are cooking spaghetti in water you have salted. You notice that when the water boils over, it causes the flame of the gas burner to turn bright orange. How would you explain the appearance of a color in the flame?

2. Write the electron arrangement for the following elements:

   N _____

   Na _____

   S _____

   Cl _____

   Ca _____

   O _____

3. Draw models of the atoms with the following isotope symbols:

   $^{18}_{8}O$

   $^{42}_{20}Ca$

   $^{14}_{6}C$

EXPERIMENT 11
BONDING OF ELEMENTS IN COMPOUNDS

GOALS

1. Identify the elements in a compound.
2. Compare some physical properties of a compound with the elements from which it was formed.
3. Use the subscripts in the formula of a compound to state the lowest whole number combination of the elements in that compound.
4. Identify the bonding in a compound as ionic or covalent.

MATERIALS NEEDED

display of elements
display of compounds
samples of iron (Fe), sulfur (S), iron and sulfur mixture (Fe + S) and iron (II) sulfide (FeS)
magnet
test tubes
dropper
6M HCl *Careful! Corrosive*

CONCEPTS TO REVIEW

formulas
ions
ionic bonds
covalent bonds
compounds

BACKGROUND DISCUSSION

Formulas

Almost everything you see around you is made of compounds. A compound is made up of at least 2 different elements that have combined by way of chemical reactions. Although there are only 109 elements, there are millions of different compounds.

EXPERIMENT 11

In any compound, there is a definite proportion of elements. A formula of a compound indicates the number of atoms of each kind of element. For example, water has the formula H_2O. This means that two atoms of hydrogen and one atom of oxygen are combined in every molecule of water. Water never has any other formula.

Types of Bonds in Compounds

Atoms form compounds to gain additional stability, usually in the form of octets in the outer shells. The resulting attractions between the atoms are called *chemical bonds*. Ionic bonds are the attractive forces between the positive ions produced by a metal and the negative ions from a nonmetal. In molecules with covalent bonds, atoms of two nonmetals share electrons to form stable outer shells.

Compound	Characteristics	Type of Bonding
NaCl	ions, (Na^+, Cl^-)	ionic
CCl_4	molecules, (C-Cl bonds)	covalent
$MgBr_2$	ions, (Mg^{2+}, Br^-)	ionic
NH_3	molecules, (N-H bonds)	covalent

LABORATORY ACTIVITIES

A. COMPOUNDS AND THEIR ELEMENTS

Wear your laboratory goggles!

Observe some compounds in the laboratory display of compounds. Write their formulas. Describe the appearance of each compound.

From the formula of each compound, state the number of atoms of each element that are combined in that compound. For example, a molecule of H_2O is composed of 2 atoms of hydrogen and 1 atom of oxygen. Observe and record some of the physical properties of those individual elements.

B. PHYSICAL PROPERTIES OF A COMPOUND AND ITS ELEMENTS
(Optional Instructor Demonstration)

Your instructor may wish to do this part of the experiment as a demonstration with test tubes already prepared. Four test tubes labeled A, B, C and D contain **small amounts** of the following substances.

Test tube	Contents
A	Fe
B	S
C	Fe + S (mixture)
D	FeS

BONDING OF ELEMENTS IN COMPOUNDS

B.1 Describe the physical appearance of the contents of each test tube.

B.2 Run a magnet along the side of each test tube. If there is any magnetic attraction, you will see particles follow the motion of the magnet. Record your observations. ***Do not place the magnet directly into the substance!*** The attracted particles cling to the magnet and make it difficult to clean.

B.3 *This part of the experiment has a reaction that produces H_2S gas which can be toxic in large amounts. Check with your instructor before proceeding.*

IN THE HOOD, add a <u>few</u> drops of 6M (6N) HCl to each test tube. Observe any reaction in the test tube. CAREFULLY note any odor.

CAUTION: TO SMELL A GAS, FIRST FILL YOUR LUNGS WITH FRESH AIR, THEN USE YOUR HAND TO FAN SOME OF THE VAPORS FROM THE TEST TUBE TOWARD YOU AS YOU CAREFULLY NOTE THE ODOR.

C. TYPES OF BONDS IN COMPOUNDS

Identify the type of units that make up each of the compounds listed. State whether the bonding is ionic or covalent.

Type of Elements	Bonding
metal and nonmetal	ionic (usually)
two nonmetals	covalent

EXPERIMENT 11
BONDING OF ELEMENTS IN COMPOUNDS
LABORATORY REPORT

A. COMPOUNDS AND THEIR ELEMENTS

Compound Formula	Physical properties	Elements in the Compound Number and symbol	Physical properties

Circle the correct answer to complete the following:

(a) Compounds (**look like <u>or</u> do not look like**) the elements from which they were formed.

(b) When elements form compounds, the characteristics of the compounds that are formed are (**new <u>or</u> the same as those of the elements**).

EXPERIMENT 11

B. PHYSICAL PROPERTIES OF A COMPOUND AND ITS ELEMENTS

	Physical properties	Magnetic attraction	Reaction with HCl
A			
B			
C			
D			

Identify the contents in test tubes A, B, C, and D as an element, a mixture, or a compound.

A_____

B_____

C_____

D_____

NAME_____ SECTION_____ DATE_____

C. TYPES OF BONDS IN COMPOUNDS

Compound	Units	Type of Bonding
LiBr		
CH_4		
$AlCl_3$		
SO_3		
Na_2S		

EXPERIMENT 11

QUESTIONS AND PROBLEMS

1. List the number of atoms of each kind of element in the following formulas:

Formula	Number and Kind of Atoms in the Compound
H_2O	2 atoms H and 1 atom O
$CuCl_2$	_____
Al_2S_3	_____
$FeSO_4$	_____
C_4H_{10}	_____
$C_6H_{12}O_6$	_____

2. Write formulas of the following compounds from the number of atoms given. The elements are listed in the order in which they appear in the formula.

1 atom of C and 2 atoms of O	CO_2
1 atom of N and 3 atoms of H	_____
1 atom Na and 1 atom of Cl	_____
1 atom of C and four atoms of Cl	_____
2 atoms of Fe and 3 atoms of O	_____
1 atom of Ba, 1 atom of S, 4 atoms of O	_____

3. Identify the bonding in the following compounds as ionic or covalent.

$BaCl_2$	_____
H_2O	_____
C_3H_8	_____
Li_2O	_____
PCl_3	_____
NaBr	_____

EXPERIMENT 12
WRITING FORMULAS AND
NAMES OF COMPOUNDS

GOALS

1. Determine the charge of an ion by observing its electron dot structure.
2. Write a correct formula and name of an ionic compound.
3. Write the name and formula of a covalent compound.
4. Write a correct formula and name of a compound with a polyatomic ion.

CONCEPTS TO REVIEW

formation of positive and negative ions
ionic and covalent compounds
writing formulas of ionic and covalent compounds
naming ionic and covalent compounds
naming polyatomic ions

BACKGROUND DISCUSSION

When we consider the chemical reactivity of elements, we are primarily interested in the electrons in the highest energy level called the **valence electrons**. The octet rule tells us that atoms become more stable when they have eight electrons in the valence shell. The exceptions are the elements H and He which are stable with two electrons.

Ionic Charges

When atoms in groups 1A, 2A, or 3A react with atoms in groups 5A, 6A, and 7A, they lose of gain electrons in their valence shells. We can predict what this loss or gain is by observing the initial electron dot structures. An electron dot structure is written by placing dots that represent valence electrons around the symbol of the atom. Aluminum, for example, whose electron configuration is 2,8,3 has three valence electrons and an electron dot structure of

$$\cdot \overset{\cdot}{Al} \cdot$$

The aluminum atom loses its three valence electrons to become stable. The resulting aluminum ion has an electron configuration of 2,8, an octet in the valence shell. As an ion, it acquires an ionic charge of 3+.

EXPERIMENT 12

```
        Al atom              Al³⁺ ion
        13 p⁺                13 p⁺
        13 e⁻                10 e⁻
        ─────                ─────
           0                   3+
```

Group A elements (metals) with 1, 2, or 3 electrons in their outershells *lose* their valence electrons; they form positively charged ions. Elements (nonmetals) with 5, 6, or 7 valence electrons *gain* electrons to become stable, and form negatively charged ions.

Writing Ionic Formulas

An ionic formula represents the smallest number of positive and negative ions that give a charge balance of zero (0).

Ions	Number of ions	Formula	Name
Ca^{2+}	1 needed	$CaCl_2$	calcium chloride
Cl^-	2 needed		

Variable Ionic Charges for Transition Elements

Many of the transition metals are capable of forming ions with more than one type of positive valence. We will illustrate variable valence with the ions of iron. The names of ions with variable valences must include the valence.

Ions	Names	Compounds	Names
Fe^{2+}	iron (II); ferrous ion	$FeCl_2$	iron (II) chloride or ferrous chloride
Fe^{3+}	iron (III); ferric ion	$FeCl_3$	iron (III) chloride or ferric chloride

Covalent (Molecular) Compounds

Two nonmetals from groups 4A, 5A, 6A, or 7A form compounds by sharing electrons. The resulting molecule has covalent bonds and is called a **covalent compound**. To write the formula of a covalent compound, you need to determine the number of electrons needed by each kind of atom to provide an octet. For example, nitrogen in group 5 has five valence electrons. Nitrogen atoms need 3 more electrons for an octet; they share 3 electrons.

WRITING FORMULAS AND NAMES OF COMPOUNDS

Electron Dot Structures and Names of Covalent Compounds

Formulas of covalent compounds are derived by sharing the valence electrons of the atoms until each atom acquires an octet. For example, oxygen has 6 valence electrons and needs two more, and hydrogen has 1 and needs just one. The formula that represents this combination is H_2O.

$$\begin{array}{c} \ddot{} \\ :\ddot{O}:H \\ H \end{array} \quad \text{octet for oxygen}$$

Polyatomic Ions

Polyatomic ions are groups of nonmetals with an overall charge, usually negative. They typically consist of a nonmetal such as sulfur combined with several oxygen atoms. Some examples are given below. The polyatomic ions are named by replacing the ending of the nonmetal with **-ate** or **-ite**. A complete list of the polyatomic ions you need to know is found in your text.

NO_3^-	CO_3^{2-}	PO_4^{3-}
nitrate	carbonate	phosphate
HCO_3^-	SO_4^{2-}	NO_2^-
bicarbonate or hydrogen carbonate	sulfate	nitrite

When two or more polyatomic ions are used in a formula, a parenthesis is used to enclose the ion and the subscript is written outside. *No change is made in the formula of the polyatomic ion itself.*

Ions	Formula	Name
Ca^{2+} 1 needed	$Ca(NO_3)_2$	calcium nitrate
NO_3^- 2 needed		

LABORATORY ACTIVITIES

A. IONIC CHARGES

In the laboratory record, complete the chart of electron dot structures for atoms and ions. Write the formula and name of the resulting ions including the ionic charges.
The positive ions have the same name as the element. The names of the negative ions replace the endings of their elemental names (nonmetals) with **-ide**.

EXPERIMENT 12

B. WRITING IONIC FORMULAS FROM THE IONS

In the laboratory record, write the formulas of the positive ions and the number of negative ions. Determine the number of each ion that will give a formula that has an overall charge balance of zero. (The total positive charge must equal the total negative charge.) Write the correct formula using subscripts to indicate the number of times each ion is used. The compound is named by writing the name of the positive ion following by the name of the negative ion.

C. VARIABLE IONIC CHARGES FOR TRANSITION METALS

In the laboratory record, write the formulas of compounds that contain ions from the transition elements with variable ionic charges. Give a correct name for each compound. Be sure to include the Roman numeral after the name of the positive ion to indicate its ionic charge.

D. COVALENT (MOLECULAR) MOLECULES

Write the electron dot structure for each of the elements listed. Then write electron dot structures for the covalent compounds listed. Give the name of the covalent compound. The subscripts in the formulas appear as prefixes in the name of the compound. Some prefixes are shown below:

Number of Atoms	Prefixes for covalent compounds
1	mono
2	di
3	tri
4	tetra
5	penta

E. POLYATOMIC IONS

In the laboratory record, determine the number of positive ions and negative polyatomic ions needed for a charge balance of zero. Write the formula using parentheses when necessary. Give the name of the compound using the name for the polyatomic ion.

NAME _____ SECTION _____ DATE _____

EXPERIMENT 12 WRITING FORMULAS AND NAMES OF COMPOUNDS LABORATORY REPORT

A. IONIC CHARGES

Element	Atomic Number	Electron arrangement of atom	Electron dot formula of atom	Loss or gain of electrons by atom	Ionic charge	Formula of ion	Name of ion
Sodium	11	2,8,1	Na•	2,8	1+	Sodium	Na⁺
Calcium							
Oxygen							
Chlorine							
Aluminum							
Potassium							
Sulfur							
Nitrogen							

EXPERIMENT 12

B. WRITING IONIC FORMULAS FROM THE IONS

Name of ion	Formula of ion	Number of ions needed	Formula of ionic compound	Name
sodium	___	___		
chloride	___	___	___	___
lithium	___	___		
oxide	___	___	___	___
magnesium	___	___		
chloride	___	___	___	___
calcium	___	___		
oxide	___	___	___	___
calcium	___	___		
nitride	___	___	___	___
aluminum	___	___		
bromide	___	___	___	___
aluminum	___	___		
oxide	___	___	___	___
sodium	___	___		
phosphide	___	___	___	___
aluminum	___	___		
nitride	___	___	___	___

C. VARIABLE IONIC CHARGES FOR TRANSITION METALS

Formula of ion	Number of ions needed	Formula of ionic compound	Name
Fe^{2+}	_____		
Br^-	_____	_____	_____
Fe^{2+}	_____		
O^{2-}	_____	_____	_____
Cu^{2+}	_____		
Cl^-	_____	_____	_____
Cu^{2+}	_____		
S^{2-}	_____	_____	_____
Fe^{3+}	_____		
S^{2-}	_____	_____	_____
Cu^+	_____		
O^{2-}	_____	_____	_____

D. COVALENT (MOLECULAR) COMPOUNDS

Element					
hydrogen	carbon	nitrogen	oxygen	chlorine	sulfur
Electron Dot Formula					

Compound	Electron dot structure	Name
CH_4	H ·· H:C:H ·· H	methane
H_2O		water
NCl_3		
H_2S		
HCl		

NAME_____ SECTION_____ DATE_____

E. POLYATOMIC IONS

Name of ion	Formula of ion	Number of ions needed	Formula of ionic compound	Name
sodium	_____	_____		
nitrate	_____	_____	_____	_____
lithium	_____	_____		
carbonate	_____	_____	_____	_____
potassium	_____	_____		
sulfate	_____	_____	_____	_____
calcium	_____	_____		
bicarbonate	_____	_____	_____	_____
aluminum	_____	_____		
hydroxide	_____	_____	_____	_____
magnesium	_____	_____		
bicarbonate	_____	_____	_____	_____
calcium	_____	_____		
phosphate	_____	_____	_____	_____
sodium	_____	_____		
phosphate	_____	_____	_____	_____

EXPERIMENT 12

QUESTIONS AND PROBLEMS

1. Write the correct formulas for the following ions:

    sodium_____   oxide _____   calcium_____

    chloride_____   sulfate _____   iron (II) _____

2. Write the correct name of the following:

    CuO _____

    N_2O_4 _____

    $Al(NO_3)_3$ _____

    PCl_3 _____

    $FeCO_3$ _____

    Na_2S _____

    $Cu(OH)_2$ _____

    Ag_2O _____

3. a. Identify the following as ionic or covalent.
   b. Write the correct formulas.

    sodium oxide _____   _____

    potassium iodide _____   _____

    magnesium fluoride _____   _____

    iron (III) bromine _____   _____

    copper (II) chloride _____   _____

    sodium carbonate _____   _____

    aluminum nitrate _____   _____

    carbon tetrachloride _____   _____

    nitrogen tribromide _____   _____

4. Your friend wants to know what the formula $FeSO_4$ on her vitamin bottle means and what the name is. What would you tell her?

EXPERIMENT 13
MOLES AND CHEMICAL FORMULAS

GOALS

1. Calculate the molar mass of a substance.
2. Use the mole concept to convert grams to moles.
3. Experimentally determine the simplest formula of an oxide of magnesium.

MATERIALS

dried peas or beans
crucible and cover
tongs
clay triangle
iron ring and stand
Bunsen burner
magnesium ribbon

CONCEPTS TO REVIEW

mole
molar mass
calculating moles from grams

BACKGROUND DISCUSSION

Moles and Molar Mass

You probably already use certain units to represent a collection of smaller things. When you buy eggs, you purchase a dozen. How often do you actually count the eggs to see if there are 12? Most likely never, because you know that a dozen eggs is exactly 12 eggs. When you buy a ream of paper, you don't have to count each piece to know that you have 500 sheets of paper in that ream.

A chemist uses the unit called a *mole* to represent a large collection of atoms or molecules. The *molar mass* is the formula weight of the compound determined from the sum of the atomic weights on the periodic table.

EXPERIMENT 13

Substance	Example of 1 mole Quantities Molar Mass
1 mole Ca	40.1 g Ca
1 mole H_2S	34.1 g H_2S
1 mole $Mg(NO_3)_2$	148.3 g $Mg(NO_3)_2$

In the first part of this experiment, you will develop the idea that by weighing out a mole of a substance you are also counting the number of atoms or molecules of that substance. Then you will observe 1 mole quantities of elements and compounds.

In the second part of the experiment, you will burn a piece of magnesium in air. The oxygen in the air reacts with magnesium to produce magnesium oxide.

$$\text{magnesium} + \text{oxygen} = \text{magnesium oxide}$$

The simplest (or empirical) formula of the product is determined by calculating the number of moles of magnesium and the number of moles of oxygen in the product. The mass of the magnesium is converted to moles using its molar mass, 24.3 g/mole.

$$\text{mole Mg} = \text{mass (g) magnesium ribbon} \times \frac{1 \text{ mole Mg}}{24.3 \text{ g Mg}}$$

The mass of oxygen that reacts is found by subtracting the mass of the magnesium from the total mass of the product. The number of moles of oxygen is found using its molar mass of 16.0 g/mole.

$$\text{mole O} = \text{mass (g) of oxygen reacted} \times \frac{1 \text{ mole O}}{16.0 \text{ g O}}$$

To determine the simplest formula of the product, the larger number of moles is divided by the smaller number of moles. Some examples of finding simplest formulas are given below:

For 0.080 mole Cl (larger number of moles)
and 0.040 mole Zn (smaller number)

$$\frac{0.080 \text{ mole Cl}}{0.040 \text{ mole Zn}} = \frac{2.0 \text{ moles Cl}}{1 \text{ mole Zn}} = ZnCl_2$$

For 0.150 mole Cu and 0.150 mole S (equal number of moles)

$$\frac{0.150 \text{ mole Cu}}{0.150 \text{ mole S}} = \frac{1.0 \text{ mole Cu}}{1 \text{ mole S}} = CuS$$

MOLES AND CHEMICAL FORMULAS

LABORATORY ACTIVITIES

A. COUNTING ATOMS

A.1 Using the laboratory balance, find the mass of 10 dried peas or beans.

A.2 Calculate the average mass.

$$\frac{\text{total mass}}{10} = \text{average mass (g)/pea or bean}$$

A.3 Calculate the mass that would contain 50 peas or beans. *Do not count them*.

A.4 Weigh out the quantity you calculated. *Now count the actual number.* Record.

A.5 Calculate the mass in grams that would contain 5000 peas or beans. *Don't try to count them!* As the number becomes larger, you will see that it becomes difficult to count them individually.

B. MOLES AND MOLAR MASS

In the laboratory, observe the 1 mole quantities of several substances. Write down the formula of the compound. Calculate the mass of compound that is present in each of the samples by determining the molar mass of each.

C. FINDING THE SIMPLEST FORMULA

C.1 Obtain a clean, dry crucible, and cover. (A porcelain crucible may have stains in the porcelain that cannot be removed.) Weigh them to the nearest 0.01 g.

C.2 Obtain a length of magnesium ribbon that has a mass of about 0.5 g. If the ribbon is tarnished, the dark oxide coating can be removed by polishing the ribbon with steel wool. Wind the ribbon into a loose coil and place in the bottom of the crucible. Replace the cover. Weigh the ribbon, crucible, and cover to the nearest 0.01 g.

Place the crucible and ribbon in the clay triangle. Place the cover nearby along with a pair of tongs. Begin heating the crucible. Recall that the tip of the inner flame should touch the bottom of the crucible.

 As soon as the magnesium begins to smoke or bursts into flame, place the cover on the crucible using your tongs. Avoid looking directly at the bright flame of the burning magnesium.

Every minute or two, use the tongs to lift up the cover of the crucible. When the magnesium *no longer* produces smoke or glows brightly, set the cover ajar and heat strongly for 5 more minutes. Then let the crucible and its contents cool for 10 minutes. See Figure 13.1.

EXPERIMENT 13

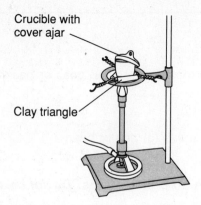

Figure 13.1 *Crucible and cover on clay triangle set on an iron ring.*

At this point, the product in the crucible should be a whitish-gray ash. Since there is also nitrogen in the air, some of the magnesium will react with the nitrogen. To get rid of the nitrogen product, use a dropper to add carefully 15-20 drops of water to the cooled contents. Return the crucible, contents and cover to the clay triangle. Set the cover slightly ajar and heat gently to avoid splattering the water, and then heat strongly for another 5 minutes. **Caution:** *ammonia which gives an irritating odor is released.*

C.3 Place the crucible, its contents, and cover on a heat resistant pad and let cool. Weigh them to the nearest 0.01 g.

Calculations

C.4 Calculate the mass of the magnesium ribbon (C.2 - C.1).

C.5 Calculate the mass of the magnesium oxide product (C.3 - C.1).

C.6 Calculate the mass of oxygen that reacted with the magnesium ribbon (C.5 - C.4).

C.7 Calculate the moles of magnesium.

mole Mg = mass(g) magnesium ribbon X $\dfrac{1 \text{ mole Mg}}{24.3 \text{ g Mg}}$

C.8 Calculate the moles of oxygen that reacted with the magnesium ribbon.

mole O = mass(g) of oxygen reacted X $\dfrac{1 \text{ mole O}}{16.0 \text{ g O}}$

C.9 Identify the larger and smaller number of moles.

C.10 Divide the larger number of moles by the smaller number of moles.

C.11 Usually the mole ratio should be small numbers such as 1:1, 1:2, 2:1, 1:3, 2:3, and so on. Pick the nearest whole number relationship to your ratio in B.10 and write the simplest formula for the oxide product.

NAME_____ SECTION_____ DATE_____

EXPERIMENT 13
MOLES AND CHEMICAL FORMULAS
LABORATORY REPORT

A. COUNTING ATOMS

A.1 Mass of 10 peas or beans   _____ g

A.2 Average mass   _____ g
     Show calculations

A.3 Mass that contains 50 peas or beans _____ g
     Show calculations

A.4 Actual number present   _____

A.5 Mass that contains 5000   _____ g

What do you think is meant by the phrase, "counting by weighing"?

B. MOLES AND MOLAR MASS

Name	Formula	Molar mass
_____	_____	_____
_____	_____	_____
_____	_____	_____
_____	_____	_____
_____	_____	_____

EXPERIMENT 13

C. FINDING THE SIMPLEST FORMULA

C.1 Mass of empty crucible + cover _____ g

C.2 Mass of crucible, magnesium + cover _____ g

C.3 Mass of crucible, product + cover _____ g

Calculations

C.4 Mass of magnesium _____ g Mg

C.5 Mass of magnesium oxide product _____ g

C.6 Mass of oxygen in the product _____ g O

C.7 Moles of magnesium _____ moles Mg
     Show calculations

C.8 Moles of oxygen _____ moles O
     Show calculations

C.9 Larger number of moles _____ moles of _____
                                                (element)

     Smaller number of moles _____ moles of _____
                                                (element)

C.10 Calculation of mole ratio:

      $\dfrac{\text{larger number of moles}}{\text{smaller number of moles}}$ = _____ mole ____ : 1 mole ____

C.11 Simplest formula _____

116

NAME_____ SECTION_____ DATE_____

QUESTIONS AND PROBLEMS

1. Calculate the molar mass for the following:

   $FeCl_3$

   $Ba(NO_3)_2$

2. Using your rules for writing the formulas of ionic compounds, write the correct formula for magnesium oxide.

3. Write a balanced equation for the reaction of magnesium with oxygen (O_2) in the air.

4. What changes in properties provide evidence of a chemical reaction?

5. Balance the following equation for the reaction of magnesium nitride to magnesium oxide.

   $Mg_3N_2(s) + H_2O(l) \longrightarrow MgO(s) + NH_3(g)$

6. Write the simplest formula for each of the following compounds:

   0.0200 mole Al and 0.600 mole Cl

   0.080 mole Ba, 0.080 mole S, 0.320 mole O

EXPERIMENT 14
PHYSICAL AND CHEMICAL CHANGES

GOALS

1. Observe physical and chemical properties associated with physical and chemical changes.
2. Use a chemistry handbook to obtain information on some physical properties.
2. Give evidence for the occurrence of a chemical reaction.
3. Write a balanced equation for a chemical reaction.

MATERIALS

test tubes
test tube rack
magnesium ribbon
zinc (s)
$CuSO_4 \cdot 5H_2O$ (s)
NH_4NO_3 (s)
$CaCl_2$ (anhydrous)
ice
thermometer
glass stirring rod

0.1 M $BaCl_2$
0.1 M Na_2SO_4
0.1 M $CuSO_4$
0.1 M $CaCl_2$
0.1 M Na_3PO_4
1 M HCl
0.1 M $FeCl_3$
1 M Na_2CO_3
1 M NH_4OH

CONCEPTS TO REVIEW

physical and chemical properties
physical and chemical changes
writing chemical equations
balancing equations

BACKGROUND DISCUSSION

This is an experiment in making observations. First, you will describe the physical properties of substances before they undergo reaction. Some physical properties are color, physical state, density, melting and boiling points. When a reaction occurs, changes occur in the properties of the substances. If a reaction causes an alteration in the properties of the reactants to give new substances, then a chemical change has occurred. A physical change occurs if the nature of the substance is retained. For example, gold can be melted. When it cools, it returns to a solid state with the same physical properties as before. When a change in temperature causes a change in state that can be reversed, the substance has undergone a physical change.

EXPERIMENT 14

Physical Changes	Chemical Changes
change in state	formation of a gas (bubbles)
change in size	formation of a solid (precipitate)
tearing	disappearance of a solid (dissolving)
breaking	change in color
grinding	evolution or absorption of heat
	a change in the pH (acidity)
	a change in the temperature as a result of a reaction

LABORATORY ACTIVITIES

 Safety glasses must be worn during this experiment

For all of the following experiments, use small quantities. For solids, use the amount of compound that will fit on the tip of a spatula or small scoop. Carefully pour small amounts of liquids into your own beakers and other containers. Measure out 2-3 mL of water in a test tube similar to those you will be using. Use this volume as a reference level. Judging by eye, use about the same level of solution for each of the experiment.

 DO NOT PLACE DROPPERS OR STIRRING RODS INTO ANY REAGENT BOTTLES. DIRTY GLASSWARE CONTAMINATES A REAGENT FOR THE ENTIRE CLASS. NEVER RETURN UNUSED CHEMICALS TO THEIR ORIGINAL CONTAINERS. THEY MUST BE DISCARDED IN THE APPROPRIATE MANNER AS INDICATED BY YOUR INSTRUCTOR.

Read directions carefully. Match the labels on bottles and containers to the names and concentrations of the substances called for in the directions. Keep your desk neat and orderly. Label each of your containers with the formulas of the chemicals as you remove them from the original laboratory bottle. (Many test tubes have a rough place for pencil writing.)

Record your observations of the physical properties of the substance before and after any reaction for each of the following experiments on the laboratory record. Carry out the reaction, and record your observations. *Use complete sentences to describe your observations.* State whether a reaction represents a physical or chemical change. Balance the equation given for the reaction in the laboratory record.

A. BURNING MAGNESIUM

Obtain a small strip of magnesium ribbon. Record your observations of its physical properties. Holding the end of the strip using a pair of <u>tongs</u>, place the magnesium ribbon in the flame of the Bunsen burner. **Shield your eyes** when the ribbon ignites and remove the ribbon from the flame. Record your observations of the reaction and the physical properties of the produce. Balance the equation for the reaction.

$$Mg\ (s)\ +\ O_2\ (g)\ \longrightarrow\ MgO\ (s)$$

B. MELTING ICE

Place an ice cube or some ice chips on a watch glass. Describe the physical properties of the ice. Record your observations again at 10 minutes, and 30 minutes. Describes the changes you observe as physical or chemical.

$$H_2O\ (s) \longrightarrow H_2O\ (l)$$

C. ZINC AND COPPER (II) SULFATE

Place 2-3 mL of 0.1 M $CuSO_4$ *solution* into each of two test tubes. Describe the appearance of the solutions and a small piece of zinc metal. Add the Zn metal to the first test tube. Keep the second test tube as a reference for color. Observe the color of the $CuSO_4$ solutions in both test tubes, and the color and appearance of the zinc in the first test tube. Repeat your observations at 10 minutes and 30 minutes.

$$Zn(s) + CuSO_4(aq) \longrightarrow Cu(s) + ZnSO_4(aq)$$

D. EXOTHERMIC AND ENDOTHERMIC REACTIONS

When heat is given off by a reaction (temperature increase), the reaction is exothermic. Heat is written as a product in the equation. When the temperature drops, heat is absorbed, and the reaction is endothermic. Heat is written as one of the reactants.

D.1 Place 5 mL of water in a test tube. Using a thermometer, measure its temperature. Record. Add a scoop of NH_4NO_3 (s) crystals to the water. Stir and record the temperature of the solution. In an equation, the state of a substance may be indicated by (s) for solid, liquid by (l), gas by (g), and dissolved in water (aqueous) by (aq). Determine whether the reaction is exothermic or endothermic.

$$NH_4NO_3(s) \xrightarrow{H_2O} NH_4^+(aq) + NO_3^-(aq)$$

D.2 Place 5 mL of water in another test tube. Measure the temperature of the water. Add a scoop of *anhydrous* $CaCl_2$ (s). Anhydrous means *without water*. Stir and record the temperature of the resulting solution.

$$CaCl_2(s) \xrightarrow{H_2O} Ca^{2+}(aq) + Cl^-(aq)$$

E. FORMATION OF PRECIPITATES (SOLID PARTICLES) AND GASES

In the following experiments, two solutions will be combined. Observe each of the substances before mixing, and again after mixing. Look for changes in color, formation of a solid called a precipitate (solution turns cloudy), dissolving of a solid, and formation of a gas (bubbling). Use 2-3 mL of each of the 0.1 M solutions. Balance the equations for the reactions.

EXPERIMENT 14

E.1 Obtain 2-3 mL each of 0.1 M $CaCl_2$ and 0.1 M Na_3PO_4. Describe each and mix together. Describe the changes that occur.

$$CaCl_2 + Na_3PO_4 \longrightarrow Ca_3(PO_4)_2(s) + NaCl$$

E.2 Obtain 2-3 mL each of 0.1 M $BaCl_2$ and 0.1 M Na_2SO_4

$$BaCl_2 + Na_2SO_4 \longrightarrow BaSO_4(s) + NaCl$$

E.3 Place 2-3 mL of 0.1 M $FeCl_3$ in a test tube and 2-3 mL of 1 M NH_4OH

CAUTION: BE CAREFUL WHEN YOU POUR OUT AMMONIUM HYDROXIDE, THE AMMONIA ODOR CAN BE IRRITATING.

$$FeCl_3 + NH_4OH \longrightarrow Fe(OH)_3(s) + NH_4Cl$$

E.4 Place 2-3 mL of 1 M Na_2CO_3 in one test tube and 2-3 of 1 M HCl in another.

Caution: HCl is corrosive. Any spills must be cleaned up immediately. If spilled on the skin, wash continuously with water for 10 minutes.

Observe any changes at the time you mix the solutions.

$$Na_2CO_3 + HCl \longrightarrow CO_2(g) + H_2O + NaCl$$

NAME_____ SECTION_____ DATE_____

EXPERIMENT 14
CHEMICAL AND PHYSICAL CHANGES
LABORATORY REPORT

A. BURNING MAGNESIUM

Physical properties	
Before reaction	After reaction
Type of change:	

____ Mg(s) + ____ O_2(g) ⟶ ____ MgO(s)

B.   MELTING ICE

Physical properties		
Initially	At 10 minutes	At 30 minutes
Type of change:		

____ H_2O(s) ⟶ ____ H_2O(l)

EXPERIMENT 14

C. ZINC AND COPPER (II) SULFATE

Physical Properties	
Before Reaction	After Reaction
Type of change:	

____Zn(s) + ____$CuSO_4$ ⟶ ____Cu(s) + ____$ZnSO_4$

D. EXOTHERMIC AND ENDOTHERMIC REACTIONS
D.1

Physical Properties	
Before Reaction	After Reaction
Type of change:	

____NH_4NO_3(s) $\xrightarrow{H_2O}$ ____NH_4^+(aq) + ____NO_3^-(aq)

D.2

Physical Properties	
Before Reaction	After Reaction
Type of change:	

____$CaCl_2$ (s) $\xrightarrow{H_2O}$ ____Ca^{2+} (aq) + ____Cl^- (aq)

NAME_____ SECTION_____ DATE_____

E. FORMATION OF PRECIPITATES (SOLID PARTICLES) AND GASES

E.1

Physical Properties	
Before Reaction	After Reaction
Type of change:	

____$CaCl_2$ + ____Na_3PO_4 ⟶ ____$Ca_3(PO_4)_2(s)$ + ____$NaCl$

E.2

Physical Properties	
Before Reaction	After Reaction
Type of change:	

____$BaCl_2$ + ____Na_2SO_4 ⟶ ____$BaSO_4(s)$ + ____$NaCl$

EXPERIMENT 14

E.3

Physical Properties	
Before Reaction	After Reaction
Type of change:	

___FeCl$_3$ + ___NH$_4$OH ⟶ ___Fe(OH)$_3$(s) + ___NH$_4$Cl

E.4

Physical Properties	
Before Reaction	After Reaction
Type of change:	

___Na$_2$CO$_3$(aq) + ___HCl(aq) ⟶ ___CO$_2$(g) + ___H$_2$O + ___NaCl

NAME_____ SECTION_____ DATE_____

QUESTIONS AND PROBLEMS
1. Using a chemistry handbook, give the physical properties of the following:

Substance	Color	Density	Melting point	Boiling point
NaCl	_____	_____	_____	_____
Pb (lead)	_____	_____	_____	_____
CaI_2	_____	_____	_____	_____
CrF_2	_____	_____	_____	_____

2. What evidence of a chemical reaction might you see in the following?

   a. fizzing of an Alka-Seltzer tablet dropped in a glass of water

   b. bleaching a stain

   c. burning a match

   d. rusting of an iron nail

3. Balance the following equations:

   a. ___Mg (s) + ___HCl $\longrightarrow$ ___H_2 (g) + ___$MgCl_2$

   b. ___Al (s) + ___O_2 (g) $\longrightarrow$ ___Al_2O_3 (s)

   c. ___Fe_2O_3 + ___H_2O $\longrightarrow$ ___Fe$(OH)_3$

   d. ___Ca$(OH)_2$ + ___HNO_3 $\longrightarrow$ ___Ca$(NO_3)_2$ + ___H_2O

4. Write an equation for the following reactions. Remember that gases of elements such as oxygen are diatomic, O_2. Balance each equation.

   a. potassium and oxygen gas react to form potassium oxide

   b. sodium and water react to form sodium hydroxide and hydrogen gas

   c. iron and oxygen gas react to form iron (III) oxide.

EXPERIMENT 15
FORMULA OF A HYDRATED SALT

GOALS

1. Determine the percent water in a hydrated salt.
2. Calculate the number of moles of water in a hydrated salt.
3. Determine the formula of a hydrated salt.

MATERIALS NEEDED

$CuSO_4 \cdot 5H_2O$
unknown: hydrated $MgSO_4 \cdot ?H_2O$
unknown: hydrated $BaCl_2 \cdot ?H_2O$
crucible
clay triangle
tongs
iron ring and stand
Bunsen burner

CONCEPTS TO REVIEW

formula
hydrates
dehydration

BACKGROUND DISCUSSION

Many solid forms of salts contain a specific number of water molecules. These compounds are called *hydrates*. The molecules of water contained in the hydrate are referred to as *water of hydration*.

Since the number of water molecules is specific for each kind of hydrate, water is included in the formula of the compound. A hydrate formula is the formula of the salt followed by a dot and the number of water molecules, as shown by the following:

$$CaSO_4 \cdot 2H_2O \qquad CuSO_4 \cdot 5H_2O \qquad Na_2CO_3 \cdot 10H_2O$$

Since the water molecules are held by weaker attractive forces than the ionic bonds of the compound, the water can be removed by heating the hydrate. The form of the salt without water is called an *anhydrate*.

$$\underset{\text{hydrate}}{CuSO_4 \cdot 5H_2O} \xrightarrow{\text{heat}} \underset{\text{anhydrate}}{CuSO_4} + \underset{\text{water of hydration}}{5H_2O\ (g)}$$

EXPERIMENT 15

By weighing the hydrate before and after heating, the amount of water lost can be calculated. The percent water in the hydrate is then determined.

$$\text{g salt hydrate} - \text{g salt anhydrate} = \text{g water}$$

$$\frac{\text{g water}}{\text{g hydrate}} \times 100 = \text{\% water in hydrate}$$

The formula of the hydrate is determined by calculating the moles of remaining compound (anhydrate) and the moles of water.

$$\text{g water} \times \frac{1 \text{ mole}}{18.0 \text{ g water}} = \text{moles water}$$

$$\text{g salt (anhydrate)} \times \frac{1 \text{ mole}}{\text{formula weight salt}} = \text{moles salt (anhydrate)}$$

When the moles of water is divided by the moles of salt (anhydrate), the ratio of moles of water to 1 mole of salt is obtained. This is written as the hydrate formula.

LABORATORY ACTIVITIES

A. COPPER (II) SULFATE·5H$_2$O, A HYDRATED SALT

A.1   Place a small amount of the solid crystals of $CuSO_4 \cdot 5H_2O$ (s) in a test tube. Describe the $CuSO_4 \cdot 5H_2O$ crystals. Place a test tube holder on the test tube. Gently heat the lower portion of the test tube by continuously moving the test tube through the flame. The proper way to heat a substance in a test tube is shown in Figure 15.1.

 DO NOT HEAT A TEST TUBE IN ONE SPOT. KEEP THE TEST TUBE MOVING.

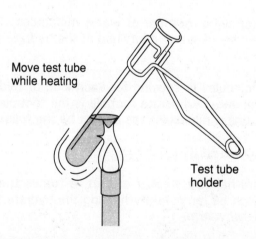

Figure 15.1 *Heating a sample in a test tube.*

FORMULA OF A HYDRATED SALT

A.2   Describe the appearance of the $CuSO_4 \cdot 5H_2O$ crystals after heating. Also observe the upper part of the test tube and describe what you see. Return the test tube to your test tube rack and let it cool. Balance the equation for the reaction. In the hydrate formula the 5 molecules of water attached follow the central dot.

$$CuSO_4 \cdot 5H_2O \text{ (s)} \longrightarrow CuSO_4 \text{ (s)} + H_2O \text{ (g)}$$

A.3   When the test tube and contents have cooled, add a few drops of water to the test tube. Record your observations.

$$CuSO_4 \text{ (s)} + H_2O \text{ (l)} \longrightarrow CuSO_4 \cdot 5H_2O \text{ (s)}$$

B. DEHYDRATION

Safety goggles are required for laboratory experiments!

B.1   Obtain a clean, dry crucible and weigh it to the nearest 0.01 g.

B.2   Place about 2 g of magnesium sulfate ($MgSO_4$) hydrate or barium chloride ($BaCl_2$) hydrate in the crucible. Weigh the crucible and its contents to 0.01 g. Record the mass.

B.3   Set the crucible and hydrate on a clay triangle that is set on a iron ring. Heat gently for 5 minutes. (See Figure 13.1 in experiment 13 for the proper setup.) Then increase the intensity of the flames and heat strongly for 10 minutes more. The bottom of the crucible will turn a dull red. Check for water vapor by holding a beaker of cold water above the crucible. If condensation of water occurs on the glass, continue to heat 5 more minutes. If no condensation occurs, turn off the burner, and let the crucible cool.

Always allow heated items to cool to room temperature. Hot containers must never be placed on a balance pan.

After cooling, weigh the crucible and its contents to the nearest 0.01 g. Record the mass.

You may wish to reheat the crucible and contents to make sure that you have completely driven off all of the water of hydration. Reheat the crucible and the anhydrate for another 5 minutes and cool again. Reweigh the crucible and contents. If the mass in the second heating is within 0.05 g of the mass obtained after the first heating, you have dehydrated the salt. If not, reheat again until you have agreement between final weighings. Use the *final weighing* for your calculations.

Calculations

B.4   Calculate the mass of the salt before heating (B.2-B.1).

B.5   Calculate the mass of the salt anhydrate after heating (B.3-B.1).

EXPERIMENT 15

B.6 The difference in the mass of the hydrate and the anhydrate is the mass of water lost during the heating. Calculate the grams of water lost (B.4 - B.5).

B.7 Calculate the percent water in the hydrate [(B.6 ÷ B.4) x 100].

$$\frac{\text{g water}}{\text{g hydrate}} \times 100 = \text{\% water in hydrate}$$

B.8 Calculate the moles of H$_2$O lost by the hydrate (B.6 ÷ 18.0 g/mole).

$$\text{g water} \times \frac{1 \text{ mole}}{18.0 \text{ g water}} = \text{moles water}$$

B.9 Calculate the moles of salt anhydrate in the sample (B.5 ÷ molar mass).

The molar mass of MgSO$_4$ is 120.4 g/mole.
The molar mass of BaCl$_2$ is 208.3 g/mole.

$$\text{g salt (anhydrate)} \times \frac{1 \text{ mole}}{\text{g in 1 mole of salt}} = \text{moles salt (anhydrate)}$$

B.10 Determine the ratio of moles of water in 1 mole of salt (B.8 ÷ B.9).

$$\frac{\underline{}\text{moles water}}{\underline{}\text{moles salt}} = \underline{}\text{moles H}_2\text{O : 1 mole salt}$$

B.11 Round off the value in A.10 to its nearest whole number. Use that number to replace the question mark in the formula of the hydrate you used. Write the correct hydrate formula for the hydrated salt used in your experiment.

MgSO$_4$·?H$_2$O

BaCl$_2$·?H$_2$O

NAME_____ SECTION_____ DATE_____

EXPERIMENT 15
FORMULA OF A HYDRATED SALT
LABORATORY REPORT

A. HEATING COPPER (II) SULFATE · $5H_2O$, A HYDRATED SALT

Physical properties	
A.1 Before heating	A.2 After heating
Type of change:	

___$CuSO_4 \cdot 5H_2O$ (s) $\xrightarrow{\text{heat}}$ ___$CuSO_4$ (s) + ___H_2O (g)

A.3   Effects of adding water:

___$CuSO_4$ (s) + ___H_2O ⟶ ___$CuSO_4 \cdot 5H_2O$ (s)

B. DEHYDRATION

Hydrate used (check one)

    ____$MgSO_4 \cdot ?H_2O$ other _____
                                       (formula)
    ____$BaCl_2 \cdot ?H_2O$

Data:

B.1   Mass of crucible _____ g

B.2   Mass of crucible and salt (hydrate) _____ g

EXPERIMENT 15

B.3 Mass of crucible and salt (anhydrate)

   first heating _____ g

   second heating _____ g

Calculations:

B.4 Mass of salt hydrate _____ g

B.5 Mass of salt (anhydrate) _____ g

B.6 Mass of water lost _____ g

B.7 % water _____ %
     Show calculations

B.8 Moles of water _____ mole
     Show calculations

B.9 Moles of salt (anhydrate) _____ mole
     Show calculations

B.10 Ratio of moles of water to moles of hydrate

   $\dfrac{\_\_\_\_ \text{ mole water}}{\_\_\_\_ \text{ mole salt}}$ = _____ moles H$_2$O : 1 mole salt

B.11 Formula of hydrated salt _____

QUESTIONS AND PROBLEMS

1. Using the formula you obtain in B.11, write an equation for the dehydration of the hydrate.

2. Write an equation for the dehydration of the following hydrates:

   $CaCl_2 \cdot 2H_2O \longrightarrow$

   $CoCl_2 \cdot 6H_2O \longrightarrow$

3. What is the % water in the hydrate $Na_2CO_3 \cdot 10H_2O$?

4. Will the experimental %H_2O be too high or too low if the hydrate sample is not heated sufficiently to drive off all the H_2O?

EXPERIMENT 16
DETECTING RADIATION

GOALS

1. Observe the use of a Geiger-Müeller radiation detection tube.
2. Determine the effect of shielding materials, distance, and time on radiation.
3. Complete a nuclear equation.

MATERIALS NEEDED

Geiger-Müeller radiation detection tube
radioactive sources
meterstick
shielding materials such as lead, paper, glass, cardboard, etc.

CONCEPTS TO REVIEW

radioactivity
alpha particles
beta particles
gamma rays
Geiger-Müeller counter
radiation protection

BACKGROUND DISCUSSION

Radioactivity occurs when there is a change in the protons and/or neutrons of the nucleus of an unstable atom. When such a change takes place, alpha particles(α), beta particles(β), or gamma rays( ) are emitted by the nucleus. These particles and rays are collectively called *nuclear radiation*.

alpha radiation $\qquad ^{147}_{62}Sm \longrightarrow ^{143}_{60}Nd + ^{4}_{2}He$

$\qquad\qquad\qquad\qquad\qquad\qquad\qquad\qquad$ *alpha particle (α)*

beta radiation $\qquad ^{45}_{20}Ca \longrightarrow ^{45}_{21}Sc + ^{0}_{-1}e$

$\qquad\qquad\qquad\qquad\qquad\qquad\qquad\qquad$ *beta particle (β)*

gamma radiation $\qquad ^{167m}_{68}Er \longrightarrow ^{167}_{68}Er + \gamma$

$\qquad\qquad\qquad\qquad\qquad\qquad\qquad\qquad$ *gamma ray*

If radiation passes through the cells of the body, the cells may be damaged. You can protect yourself by using shielding materials, by limiting the amount of time spent in radioactive areas, and by keeping a reasonable distance from the radioactive source.

To detect radiation a device such as a Geiger-Müeller tube is used. Radiation passes through the gas held within the tube, producing ion pairs. These charged particles emit bursts of current that are converted to flashes of light and audible clicks. In this experiment your teacher will demonstrate the use of the Geiger-Müeller tube to test the effects of shielding, time, and distance.

LABORATORY ACTIVITIES

     SAFETY GOGGLES PLEASE!

(*This experiment may be done as a demonstration*)

     ***RADIATION IS HAZARDOUS:** FOLLOW YOUR INSTRUCTOR'S DIRECTIONS CAREFULLY.*

A. BACKGROUND COUNT

A.1   The level of radiation that occurs naturally is called *background radiation*. Set the radiation counter at the proper voltage for operating level. Let it warm up for 5 minutes. Remove all sources of radiation near the counter. Count any radiation present in the room by operating the counter for 1 minute. Record the counts. Repeat the background count for two more 1-minute intervals.

A.2   Total the counts in A.1, and divide by 3. This value represents the background level of radiation in counts per minute (cpm). This background count is the natural level of radiation that constantly surrounds and strikes us. For the rest of this experiment, the background radiation *must be subtracted* to give the radiation from the radioactive source alone.

B. RADIATION FROM RADIOACTIVE SOURCES

B.1   Obtain a radioactive source. Place the radioactive sample at a distance of 10 - 20 cm from the detection tube.

B.2   Record the radiation emitted for 1 minute. Subtract the background count to obtain the radiation emitted by the source alone.

DETECTING RADIATION

C. EFFECT OF SHIELDING, TIME AND DISTANCE

C.1 **Shielding** Set the same radiation source at the same distance from the detection tube as in B. Place various shielding materials (cardboard, paper, several pieces of cardboard, lead sheet, etc.) between the radioactive source and the detection tube. Do not vary the distance of the source from the tube. (See Figure 16.1) Record the type of shielding used.

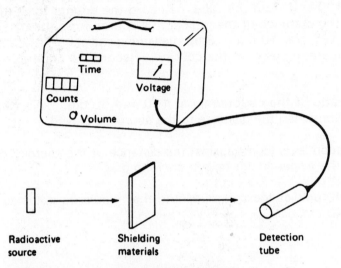

Figure 16.1 *Measuring radiation using shielding.*

C.2 Record the counts/minute obtained from the source with each type of shielding.

C.3. **Time** The greater the length of time you spend near a radioactive source, the greater the amount of radiation you receive. See Figure 16.2.

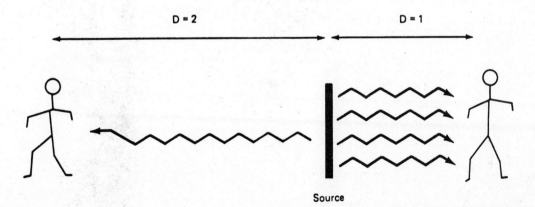

Figure 16.2 *The effect of radiation lessens as the distance from the source increases.*

139

EXPERIMENT 16

Keeping the distance constant, record the total counts for a radioactive source over a time period of 1, 2 and 5 minutes. Subtract background radiation from each to obtain the radiation from the source alone.

C.4   Use the results from C.3 to calculate the radiation that would be received after 10, 30, and 60 minutes.

C.5   **Distance**   By doubling your distance from a radioactive source, you receive one-fourth (1/4) the intensity of the radiation.   See Figure 16.2. Place a radioactive source 1 m (100 cm) from the tube.   Record the counts for 1 minute at 100 cm. Decrease the distance of the radioactive source from the detection tube to 75 cm, 50 cm, 25 cm, 10 cm.   Stop measuring radiation if the counts get too great for the operating level of the counter. Record the counts for 1 minute at each distance.

C.6   Calculate the ratio of the counts/min at 100 and 50 cm. This will give the increase in radiation when the distance to the source is halved.

C.7   Graph the radiation level (cpm) against the distance of the source from the detection tube. See the Appendix for help in drawing graphs.

C.8   Using the graph, predict the counts/minute that would be obtained at distances of 20 cm and 40 cm.

NAME_____ SECTION_____ DATE_____

EXPERIMENT 16
DETECTING RADIATION
LABORATORY REPORT

A. BACKGROUND COUNT

A.1   Counts during minute 1   _____

                    minute 2   _____

                    minute 3   _____

A.2   Total counts   _____

      Average background count   _____ counts/minute (cpm)

B. RADIATION FROM RADIOACTIVE SOURCES

B.1   Type of radioactive particles emitted _____
      (alpha particles, beta particles, gamma rays)

B.2   _____ — _____ = _____
         cpm   background   cpm for source

C. EFFECT OF SHIELDING, TIME AND DISTANCE

      Background Count   _____

Shielding

      C.1 Type of Shielding   C.2 Counts/minute

      _____   _____

      _____   _____

      _____   _____

      _____   _____

      _____   _____

Which type of shielding would offer the best protection from radiation?

EXPERIMENT 16

C.3 *Time*

Number of Minutes	Counts
1	_____
2	_____
5	_____

C.4   Counts for 10 minutes (calculated) _____

Counts for 30 minutes (calculated) _____

Counts for 60 minutes (calculated) _____

C.5   *Distance from Radioactive Source*   *Counts/minute*

100 cm	_____ cpm
75 cm	_____ cpm
50 cm	_____ cpm
25 cm	_____ cpm
10 cm	_____ cpm

C.6   Ratio: $\dfrac{\text{Counts/min at 20 cm}}{\text{Counts/min at 40 cm}}$ = _____

What happens to the intensity of the radiation when the distance from the detection tube doubles?

NAME_____ SECTION_____ DATE_____

C.7 *Graph*

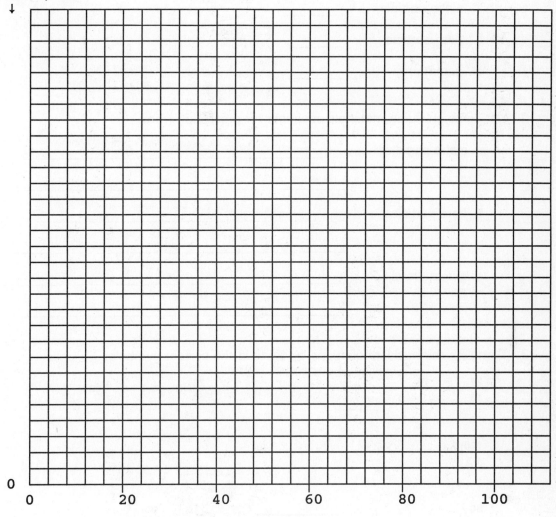

C.8   Using the graph to estimate radiation levels:

Counts/min at 40 cm = _____

Counts/min at 20 cm = _____

EXPERIMENT 16

QUESTIONS AND PROBLEMS

1. Write the symbols for of the following types of radiation:

    a. alpha particle _____

    b. beta particle _____

    c. gamma ray _____

2. List some shielding materials adequate for protection from:

    a. alpha particles _____

    b. beta particles _____

    c. gamma rays _____

3. Complete the nuclear equations by filling in the correct symbols:

    $^{27}_{13}Al\ +\ \underline{\ \ \ \ }\ \longrightarrow\ ^{24}_{11}Na\ +\ ^{4}_{2}He$

    $^{125}_{53}I\ +\ ^{0}_{-1}e\ \longrightarrow\ \underline{\ \ \ \ }$

    $^{96}_{40}Zr\ +\ \underline{\ \ \ \ }\ \longrightarrow\ ^{1}_{0}n\ +\ ^{99}_{42}Mo$

4. Describe a radioisotope used in nuclear medicine:

EXPERIMENT 17
BOYLE'S LAW

GOALS

1. Observe the effects of a pressure change upon the volume of a gas.
2. Describe the relationship between the pressure and volume of a gas.
3. Calculate the constant P x V for the experiment.
4. Graph the relationship between pressure and volume.

MATERIALS NEEDED

Boyle's law mercury apparatus

CONCEPTS TO REVIEW

gas variables
Boyle's law
inverse relationship

BACKGROUND DISCUSSION

In this experiment, you will establish the relationship between the pressure and the volume a gas. According to Boyle's law, the pressure of a gas varies inversely with the volume of the gas when the temperature and quantity are kept constant. Mathematically, this is expressed as

$$P \propto \frac{1}{V} \quad (T, n \text{ constant}) \quad \text{Boyle's law}$$

Using Boyle's law, we can say that if the volume decreases, the pressure increases. We can also rearrange Boyle's law so that the product of pressure and volume at constant temperature is constant.

$$PV = C \quad (T, n \text{ constant})$$

At constant temperature, the pressure is inversely proportional to the volume which means that the PV product remains equal. If we express initial pressure and volume as P_i and V_i, and final pressure and volume as P_f and V_f, we obtain the following equation for Boyle's law.

$$P_i V_i = P_f V_f \quad (T, n \text{ constant})$$

EXPERIMENT 17

In this experiment, you will observe changes in volume as pressure is changed at constant temperature. A PV constant will be calculated for each pressure and volume recorded. This data will be used to prepare a graph of pressure-volume relationship of Boyle's law.

Boyle's Law Apparatus

Your instructor will demonstrate the use of a Boyle's law mercury apparatus. The gas in this experiment will be a sample of air that is trapped in the closed buret in the apparatus. Initially, the buret is opened so it fills with air at atmospheric pressure. The levels of mercury in both glass tubes should be equal. At this point, the pressure of the air sample is equal to atmospheric pressure. See Figure 17.1.

Closing the stopcock causes a certain quantity and volume of air to become trapped in the closed tube. Raising or lowering the mercury reservoir changes the pressure on the air in the closed tube. As a result, the volume in the closed tube also changes. Each time the mercury reservoir is moved, a new pressure and volume are measured. The temperature and moles do not change during the experiment. The moles of air are constant because they are trapped in the buret and cannot escape. It is assumed that the temperature will not change during the experiment.

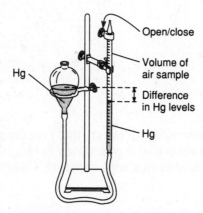

Figure 17.1 Apparatus used to measure the relationship of pressure and volume.

Measuring the pressure

When the level of mercury in the closed tube is equal to the mercury level in the open tube, the pressure of the air sample is equal to atmospheric pressure. When the level of mercury in the closed tube is lower than the level of mercury in the open tube, the pressure on the air sample is greater than atmospheric pressure.

$$P_{closed} = P_{atm} + P_{difference}$$

When the mercury level in the closed tube is higher than the mercury level in the open tube, the pressure of the air sample is lower than atmospheric pressure.

$$P_{closed} = P_{atm} - P_{difference}$$

LABORATORY ACTIVITIES

A. PRESSURE AND VOLUME OF A GAS

This experiment may be done as a demonstration.

A.1 Read a barometer to determine the atmospheric pressure during this experiment. Record that pressure in mm Hg.

A.2 Open the stopcock of the mercury apparatus and allow the levels of mercury to become equal. Then close the stopcock to trap the air sample inside the closed tube. This volume of air will be the *gas sample* for the experiment. Record the *volume*(mL) of the air sample in the closed tube.

A.3 Record the levels of mercury to the nearest millimeter (mm Hg) in the closed tube, and then in the open tube. Raise or lower the mercury reservoir to give a new pressure and volume. Record the volume (mL) of each new air sample and the mercury levels (mm Hg) in the closed and open tubes.

Calculations

A.4 For each pressure, calculate the pressure difference in mm Hg of the closed and open tube. If the mercury level in the closed tube is lower than the open tube, place a positive (+) sign in front of the difference. If the mercury level in the closed tube is higher than the open tube, place negative (—) sign in front of the pressure difference.

A.5 Add (or subtract) the pressure difference calculated in A.4 to the atmospheric pressure. This gives the pressure (mm Hg) of the gas (air) sample.

$$P_{air\ sample} = P_{atm} (+ \text{ or } -) P_{difference}$$

A.6 For each set of P and V data, obtain the product of the pressure and the volume (A.5 x A.2). Round off to the correct number of significant figures.

A.7 On the graph, the pressure (mm Hg) of the gas sample will be indicated on the vertical axis, and the volume (mL) on the horizontal axis. Use the full area of the graph paper by adjusting the lowest and highest values to fit. The lowest value of pressure should be near the lowest values obtained in the data. (You do not need to start at zero.) The volume values should begin with a value near the smallest volume attained. Be sure to mark each axis in units of pressure and volume in equal intervals. Draw a smooth line (slight curve) through the points obtained from the data.

A.8 Use the graph to discuss the meaning of Boyle's Law.

NAME_____ SECTION_____ DATE_____

EXPERIMENT 17 BOYLE'S LAW LABORATORY REPORT

Atmospheric pressure = _____ mm Hg

Reading	A.2 Volume of Air (mL)	A.3 Hg level in closed tube (mm Hg)	A.3 Hg level in open tube (mm Hg)	A.4 Difference in levels (mm Hg)	A.5 Presure of Air Sample (mm Hg)	A.6 PV Product
1						
2						
3						
4						
5						
6						

EXPERIMENT 17

A.7 Graphing Pressure and Volume in Boyle's Law

Pressure
(mm Hg)
↓

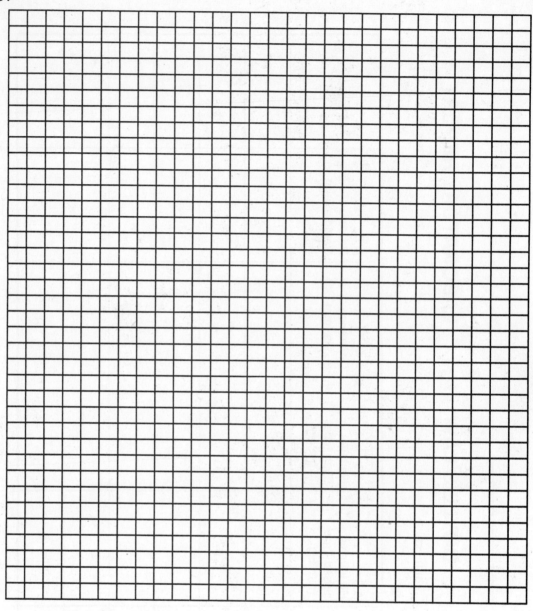

Volume of gas (mL)

A.8 According to your graph, what is the relationship between pressure and volume?

NAME_____ SECTION_____ DATE_____

QUESTIONS AND PROBLEMS

1. The behavior of a gas according to Boyle's Law requires that the temperature and number of moles of a gas be kept constant. How did this experiment keep the temperature and the moles of gas constant?

2. Why are the values for the pressure-volume product PV very close in all the readings?

3. Indicate what should happen to the pressure or volume in the following examples when T and n are constant:

Pressure	Volume
increases	_____
_____	increases
decreases	_____

4. A sample of helium has a volume of 325 mL and a pressure 650 mm Hg. What will be the pressure if the helium is compressed to 125 mL at constant temperature? (Show work.)

    _____ mm Hg

5. A 75.0 mL sample of oxygen has a pressure of 1.50 atm. What will be the new volume if the pressure becomes 45.0 atm?

    _____ mL

EXPERIMENT 18
CHARLES' LAW

GOALS

1. Observe the effect of changes in temperature upon the volume of a gas.
2. Graph the data for volume and temperature of a gas.
3. State a relationship between the temperature and volume of a gas.

MATERIALS NEEDED

400-mL beaker
graduated cylinder
water pan
thermometer
ice

The following setup may be made available by your instructor:
125 mL Erlenmeyer flask
one-hole rubber stopper fitted with a short piece of glass (polished)
400 mL beaker
wire gauze
buret clamp
rubber tubing (optional)
pinch clamp (optional)

CONCEPTS TO REVIEW

gas variables
Kelvin temperature
Celsius to Kelvin temperature equation
Charles' law

EXPERIMENT 18

BACKGROUND DISCUSSION

According to Charles' law, the volume of a gas changes directly with the Kelvin temperature as long as pressure and number of moles remain constant. This can be expressed mathematically as

$$V \propto T_K \quad (P, n \text{ constant})$$

or

$$\frac{V}{T} = C \quad (P, n \text{ constant})$$

Under two different conditions, we can write

$$\frac{V_i}{T_i} = \frac{V_f}{T_f}$$

In this experiment, we will examine changes in the volume of a gas with change in Kelvin (absolute) temperature. The volume of air contained in an Erlenmeyer flask will serve as the gas sample. A temperature of 100°C (373K) can be obtained by placing the flask in a boiling water bath. Then the volume of the gas at the boiling point of water will be the full volume of the flask. The temperature of the air contained in the flask will be lower when the flask containing trapped air is submerged in a pan of cool water. The value of absolute zero will be predicted by graphing the volume-temperature relationship and extrapolating the temperature to a point on the temperature axis when volume would be equal to zero.

LABORATORY ACTIVITIES

A. TEMPERATURE AND VOLUME OF A GAS

Wear protective safety goggles!

A.1 Dry any moisture on the inside or outside of a 125 mL Erlenmeyer flask. Gently warm the outside of the flask. Cool to room temperature. Obtain the rubber stopper assembly (or prepare) with a short piece of glass tubing (if there is a section of rubber tubing attached, obtain a pinch clamp). The stopper must fit the neck of your Erlenmeyer flask. Place in the Erlenmeyer flask.

Set a 400 mL beaker on an iron ring covered with a wire gauze. The height of the ring should be properly adjusted for your Bunsen burner. Attach a buret clamp to the neck of the flask and lower the flask into a 400 mL beaker as far into the flask as possible without touching the bottom. Attach the buret clamp to the ring stand. Add water up to the neck of the flask. Allow enough space at the top of the beaker for the water to boil but not splatter or boil over. See Figure 18.1. Begin heating and bring the water to a boil. Continue to boil (gently) for 5 more minutes to allow the air in the flask to come to the same temperature as the boiling water. Record the temperature of the boiling water.

CHARLES' LAW

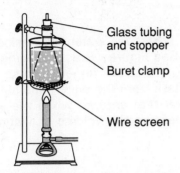

Figure 18.1 Setup for flask in a boiling water bath.

While the water is boiling, obtain a water pan or a deep container (bucket) and fill about 3/4 full with tap water. Some warmer water may be added to adjust the temperature of the water bath. Turn off the burner. Prepare to remove the flask from the hot water.

 BE CAREFUL WHILE WORKING WITH A BOILING WATER BATH. HOT WATER BURNS ARE PAINFUL.

Undo the buret clamp from the pole. Place your finger tightly over the end of the glass tubing (optional: place a pinch clamp on the rubber tubing in the stopper). Keeping your finger over the end of the tubing, *carefully* lift the flask out the hot-water using the buret clamp as a handle. While keeping the stopper end pointed downward, immerse the flask into the pan of cool water. While you continue to keep the stopper end of the flask down, remove your finger (or undo the pinch clamp). See Figure 18.2.

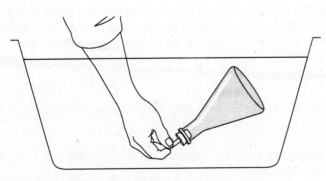

Figure 18.2 Placing the heated flask in a pan of water.

155

EXPERIMENT 18

The reduction in temperature will cause a decrease in the volume of the air in the flask, and water will enter the flask. Keep the flask submerged in the water for 10 minutes. *KEEP THE FLASK INVERTED THE ENTIRE TIME TO KEEP THE AIR TRAPPED INSIDE.*

A.2 As you prepare to remove the flask from the water pan, record the temperature of the water in the water pan.

A.3 ***Keeping the flask inverted (upside down)***, adjust the water level inside the flask to match the level of water in the pan. See Figure 18.3. Keeping the flask submerged, tightly cover the glass tubing with your finger (or use the pinch clamp on the rubber tubing). Remove the flask from the water, and set it upright on your desk. The flask will contain water that entered when the air volume decreased. Using a graduated cylinder, measure the volume of water found in the flask. Record.

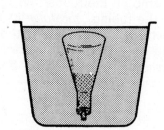 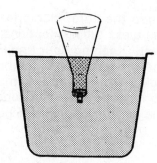

Figure 18.3 *Equalizing the water levels inside and outside the inverted flask.*

Second run

Dry and warm the flask again. Replace the stopper assembly and heat in a boiling water bath as you did in step A.1. While the water is boiling, prepare an ice-water mixture in the water pan. Repeat the earlier procedure of heating and transferring the flask. Remember to keep the flask upside down as you submerge it in the ice-water mixture. Allow the flask to remain submerged for at least 10 minutes.

A.4 When you are ready to remove the flask, record the temperature of the cold water.

A.5 Equalize the water levels as before, remove, and measure the volume of water that entered the flask. Record.

A.6 Determine the full volume of the Erlenmeyer flask by filling it with water and inserting the rubber stopper assembly. Measure the volume of water remaining. Record.

Calculations

A.7 Determine the volume of air in the flask at boiling. This is equal to the full volume of water that filled the flask in step A.6.

A.8 Determine the volume of air in the flask in cool water. Subtract the volume of water that entered the flask from the full volume of the flask (A.6 - A.3).

$$\text{volume}_{air} = \text{volume}_{full} - \text{volume}_{water\ entered}$$

A.9 Determine the volume of air in the flask at ice-water temperature (A.6 - A.5).

A.10 Convert the Celsius temperatures for each to Kelvins.

A.11 Using the Kelvin temperatures, calculate the constant which is the ratio of volume/temperature. ($A.7 \div T_K$, $A.8 \div T_K$, $A.9 \div T_K$).

A.12 Graph the volume-temperature relationship of the gas. The temperature scale runs from -400°C up to 100°C for volumes of 0 to 250 mL. Theoretically, gases would reach a temperature called **absolute zero** at a volume of 0 mL. Draw a straight line as best you can through the points on the graph. Extend the line so it goes all the way to the temperature axis where the volume is equal to zero (0). The temperature where the graph crosses the horizontal axis is your prediction for absolute zero. Record this value. See Figure 18.4.

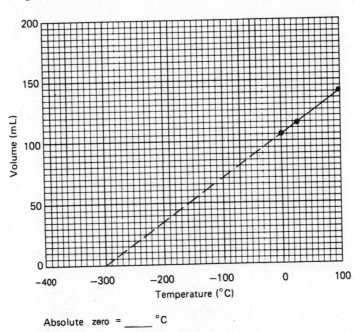

Figure 18.4 Example of graph for predicting a temperature of absolute zero.

NAME_____ SECTION_____ DATE_____

EXPERIMENT 18
CHARLES LAW
LABORATORY REPORT

A. TEMPERATURE AND VOLUME DATA

	Boiling water	Cool water	Cold water
Temperature(°C)	_____°C (A.1)	_____°C (A.2)	_____°C (A.4)
Volume of H_2O in flask	0 ml	_____mL (A.3)	_____mL (A.5)
Volume of flask		_____mL (A.6)	

Calculations

Volume of air left in flask	_____mL (A.7)	_____mL (A.8)	_____mL (A.9)
A.10 Temperature (K)	_____K	_____K	_____K
A.11 $\dfrac{\text{Volume}}{\text{Temperature}}$	_____	_____	_____

A.12 Predicted value of absolute zero _____°C

EXPERIMENT 18

A.11 Graphing Volume and Temperature Relationship

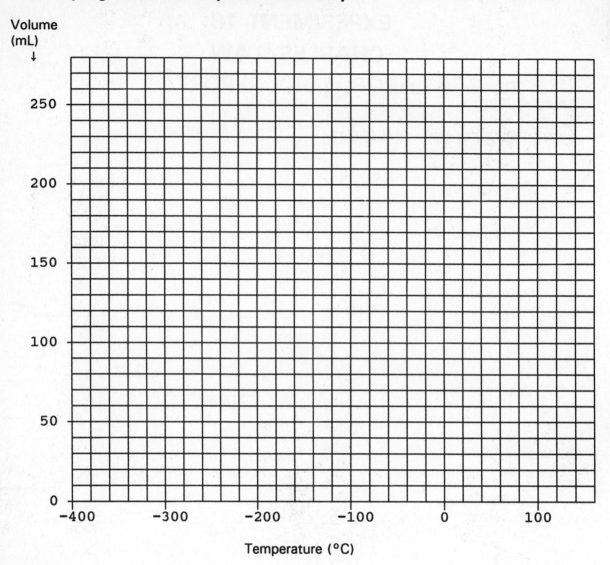

According to the graph above, what is the relationship between a change in volume and a change in temperature?

How does your predicted value for absolute temperature compare with the accepted value of -273°C?

NAME_____ SECTION_____ DATE_____

QUESTIONS AND PROBLEMS

1. Indicate the change expected when pressure is held constant for a given amount of gas.

Volume	Temperature
increases	_____
_____	decreases
decreases	_____
_____	increases

2. A gas with a volume of 525 mL at a temperature of -25°C is heated to 175°C. What is the new volume of the gas if pressure and number of moles are held constant?

3. A gas has a volume of 2.8 L at a temperature of 27°C. What temperature (°C) is needed to expand the volume to 44 L? (P and n are constant.)

4. *Combined gas law problem:* A balloon is filled with 50.0 mL of helium at a temperature of 27°C and 755 mm Hg. As the balloon rises in the atmosphere, the pressure and temperature drop. What volume will it have when it reaches an altitude where the temperature is -33°C and the pressure is 0.65 atm?

EXPERIMENT 19
PARTIAL PRESSURES OF THE GASES IN ATMOSPHERIC AND EXHALED AIR

GOALS

1. Measure the percentage of oxygen in the air.
2. Calculate the partial pressures of oxygen and nitrogen in air.
3. Measure the partial pressure of CO_2 in atmospheric air and in expired (exhaled) air.

MATERIALS NEEDED

iron filings
250 mL beaker
large test tube
graduated cylinder
marking pen or tape
shell vials
6 M NaOH
mineral oil
100 mL beaker
food coloring (optional)
meter stick

The following items may already be assembled and the setup available for your use
250-mL Erlenmeyer flask
two-holed rubber stopper fitted with two short pieces of
   glass tubing
rubber tubing (1 short; 1 long)
1 long piece of glass tubing, 60-75 cm
pinch clamps
straws to fit into end of rubber tubing

CONCEPTS TO REVIEW

partial pressures
Dalton's law
gas mixtures
atmospheric pressure
pressure gradient

EXPERIMENT 19

BACKGROUND DISCUSSION

The two major gases in the air are oxygen (O_2) and nitrogen (N_2). In this experiment, you will make measurements that will be used to calculate the percentage of oxygen in the air. The amount of oxygen in air is determined by removing the oxygen from a sample of air, and measuring the change in volume. To remove the oxygen, we place some iron filings in a moistened test tube and allow them to react.

$$4Fe + 3O_2 \longrightarrow 2Fe_2O_3$$

The oxygen is used up by the chemical reaction and the volume it occupied is replaced by water. Therefore, the volume of water that enters the container is equal to the volume of oxygen in the air sample. By measuring the volume of the air in the container, and the volume of oxygen, the percentage of oxygen in air can be calculated. If the atmospheric pressure is known, you can calculate the partial pressures (mm Hg) of oxygen gas and nitrogen gas in the air. **Partial pressures** are the individual pressures exerted by each of the gases that make up the total atmospheric pressure.

When the body metabolizes nutrients, one of the end products is carbon dioxide, CO_2. The level of carbon dioxide in the blood triggers breathing mechanisms and provides the correct pH of the blood. Accumulation of carbon dioxide above these levels can result in respiratory and metabolic dysfunction and possible death. The body eliminates most of the carbon dioxide waste by exhalation of air from the lungs.

In the second part of the experiment, you will determine the partial pressures of carbon dioxide in the atmosphere, and in expired (exhaled) air. Mineral oil placed on top of the sodium hydroxide prevents it from reacting with carbon dioxide. When the system is closed, the vial inside the flask containing the NaOH is tipped over, and the sodium hydroxide reacts with the carbon dioxide in the air samples.

$$CO_2 (g) + NaOH \longrightarrow NaHCO_3$$

The other gases in the atmosphere, oxygen and nitrogen, do not react with sodium hydroxide and continue to exert their respective partial pressures. The difference between the atmospheric pressure and the pressure of the remaining gases is used to calculate the partial pressure of CO_2 in atmospheric air and in expired air.

LABORATORY ACTIVITIES

A. PARTIAL PRESSURES OF OXYGEN AND NITROGEN IN AIR

Safety glasses are required in the laboratory!

A.1 Completely fill a large test tube with water. Measure the volume of the water in the test tube by emptying the water into a graduated cylinder. This is the initial volume of air. Record the volume of the test tube.

A.2 Obtain a small scoop of iron filings and sprinkle them in the empty, but still moist test tube. If the test tube is too dry, add a few drops of water. Shake the iron filings about the test tube. Some should adhere to the sides. Shake out the excess filings.

Fill a 250 mL beaker about one-half full of water. Place the test tube containing the iron filings upside down in the water. Place a test tube holder on the test tube and let the handle rest on the rim of the beaker. This will stabilize the inverted test tube. Carefully place the beaker and test tube in your drawer. This part of the experiment will be finished at the beginning of the next laboratory period. See Figure 19.1.

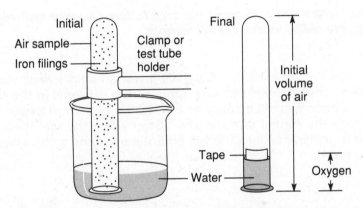

Figure 19.1 Beaker and test tube assembly for determination of the partial pressures of oxygen and nitrogen in the air.

When you return to the laboratory, you should find that water has entered the test tube. This happens because the iron filings in the test tube react with the oxygen in the air. As the oxygen is used up, it is replaced by water. Use a marking pen or tape to mark the water level inside the test tube. Then remove the test tube from the beaker.

Fill the test tube with water *up to the line* you marked. Empty the water into a graduated cylinder and record its volume. Because you inverted the test tube, this volume represents the nitrogen in the air that *remained* in the test tube after the reaction.

A.3 Read a barometer and record the barometric pressure in mm Hg.

Calculations

A.4 Subtract the volume of nitrogen measured in step A.2 from the total volume measured in step A.1. The difference in the two volumes (A.1 - A.2) is equal to the volume of oxygen in the air sample.

A.5 Calculate the percentages of oxygen and nitrogen in air. Divide each of their calculated volumes by the total volume of the air sample.

$$\frac{\text{volume (gas)}}{\text{volume (air sample)}} \times 100 = \% \text{ gas } (O_2 \text{ or } N_2) \text{ in air}$$

EXPERIMENT 19

A.6 Calculate the partial pressures of oxygen and nitrogen by multiplying each percentage by the atmospheric pressure.

$$\frac{\% \ O_2 \ (or \ N_2)}{100} \times \text{atmospheric pressure} = \text{partial pressure } O_2 \ (or \ N_2)$$

B. CARBON DIOXIDE IN THE ATMOSPHERE

This may be a demonstration by the instructor. Inquire whether a setup is already available for use.

Sodium hydroxide is caustic! Be sure to clean up any spill immediately. Wash any spills on skin for 10 minutes.

Prepare a setup as seen in Figure 19.2. Carefully lower an empty shell vial into a 250 mL Erlenmeyer flask so that the vial remains upright. Set a funnel in the flask directly above the shell vial. Obtain a small amount of 6 M NaOH in a small beaker, and *slowly* pour the NaOH through the funnel into the vial. Stop when the vial is about 3/4 full. Pour a small amount of mineral oil into the vial until the oil forms a *thin* layer (1mm) on top. Remove the funnel.

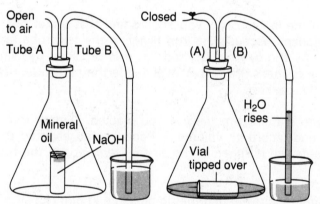

Figure 19.2 Flask and vial setup for carbon dioxide determination.

Place the two-holed stopper containing two pieces of glass tubing in the flask. Attach rubber tubing, one short (A) and one long (B), to the pieces of glass tubing. Attach a length of glass tubing (60-75 cm) to the longer piece of rubber tubing (tube B). Place the free end of the glass tubing (tube B) in a beaker that contains water and a few drops of food coloring (optional). With the rubber tubing open, the flask is full of air at atmospheric pressure. *Close* the system by attaching a pinch clamp to tube A.

B.1 Gently tilt the flask to tip over the shell vial. The NaOH solution will spill out into the bottom of the flask. Swirl or shake the flask gently. (Hold the flask around the top with your fingers. Try not to let your hands warm the flask since the increase in temperature will change the pressure.) *Make sure that tube B is*

PARTIAL PRESSURES OF GASES IN ATMOSPHERIC AND EXHALED AIR

vertical with the end below the water level in the beaker. Continue to swirl the flask while you observe changes in the water level in tube B. Wait a few minutes, then swirl again until there is no further change in the water level. Using a meter stick, measure the distance (mm) between the water level in the beaker, and the water level in the vertical glass tube B. Record.

B.2 As the carbon dioxide in the sample reacts with NaOH, there is a drop in the pressure of the gases in the sample.

$$CO_2 (g) + NaOH \longrightarrow NaHCO_3$$

As the CO_2 reacts, water will rise in the glass tube. You are observing a pressure change in mm H_2O. To convert this pressure change to mm Hg, divide the height of the water column by 13.6, the density of mercury compared to the density of H_2O. (13.6 mm H_2O exerts the same pressure as 1 mm Hg). (B.1 ÷ 13.6) This is the partial pressure of CO_2 in mm Hg for that sample of air.

$$1 \text{ mm Hg} = 13.6 \text{ mm } H_2O$$

$$\text{mm } H_2O \times \frac{1 \text{ mm Hg}}{13.6 \text{ mm } H_2O} = \text{mm Hg}$$

B.3 Read a barometer and record the atmospheric pressure in mm Hg.

Calculations

B.4 Calculate the percent CO_2 in the atmospheric air.

$$\frac{B.2}{B.3} \times 100 = \frac{\text{mm Hg } CO_2}{\text{mm Hg atm (total)}} \times 100 = \%CO_2$$

C. CARBON DIOXIDE IN EXHALED (EXPIRED) AIR

Set up the apparatus as you did in part B filling a vial with NaOH and topping with mineral oil. Fit a clean straw into the tubing of tube A. Place tube B in the beaker of water. Take a breath of air, hold for a moment, and exhale through the straw. This will cause bubbling in the beaker of water. Cover the straw with your finger while you inhale. Take a few breaths, and exhale through the straw again. Repeat this process 6 or 8 times.

Take your time. Exhaling too rapidly may cause hyperventilation and make you dizzy. Stop and rest.

EXPERIMENT 19

C.1 As you repeat the exhalations through the straw into the flask, the expired air from the lungs is replacing the atmospheric air in the flask. When you have finished the series of exhalations, close tube A with the pinch clamp. (Tube B should still be in the beaker of water.) Tip the flask to spill the NaOH from the shell vial so its can react with the CO_2 in the sample. Swirl and shake the flask gently and watch for the water to rise inside the glass tubing(B). There should be a dramatic rise in the water level inside the glass tube. If not, check your system for leaks and repeat. When no further change occur, use a meter stick to measure the distance (mm water) between the water level in the beaker and the height of the water in tube B. Record.

C.2 Convert the height of the water column to mm Hg (mm H_2O ÷ 13.6). This is the partial pressure of CO_2 in mm Hg for the expired air sample.

C.3 Record the atmospheric pressure in mm Hg.

C.4 Calculate the percentage CO_2 in expired air.

$$\frac{C.2}{C.3} \times 100 = \frac{\text{mm Hg } CO_2}{\text{mm Hg atm (total)}} \times 100 = \% CO_2$$

EXPERIMENT 19
PARTIAL PRESSURES OF GASES IN ATMOSPHERIC AND EXHALED AIR
LABORATORY REPORT

A. PARTIAL PRESSURES OF OXYGEN AND NITROGEN IN AIR

A.1 Volume of test tube _____ mL

A.2 Volume of nitrogen _____ mL

A.3 Atmospheric pressure _____ mm Hg

Calculations:

A.4 Volume of oxygen _____ mL
Show calculations:

A.5 Percent oxygen _____ %O_2
Show calculations:

Percent nitrogen _____ %N_2
Show calculations:

A.6 Partial pressure of oxygen _____ mm Hg
Show calculations:

Partial pressure of nitrogen _____ mm Hg
Show calculations:

EXPERIMENT 19

B. CARBON DIOXIDE IN THE ATMOSPHERE

B.1   Height of water column   _____mm H_2O

B.2   Height of a mercury column   _____mm Hg

    ____mm H_2O x $\dfrac{1 \text{ mm Hg}}{13.6 \text{ mm } H_2O}$ = mm Hg

B.3   Atmospheric pressure   _____mm Hg

B.4   Percent CO_2 (atmosphere)   _____% CO_2
    Show calculations:

C. CARBON DIOXIDE IN EXHALED (EXPIRED) AIR

C.1   Height of water column   _____mm H_2O

C.2   Height of a mercury column   _____mm Hg
    Show calculations:

C.3   Atmospheric pressure   _____mm Hg

C.4   Percent CO_2 (expired air)   _____% CO_2
    Show calculations:

NAME_____ SECTION_____ DATE_____

QUESTIONS AND PROBLEMS

1. In part A, what would happen to your calculated value of the %O_2 in the atmosphere if some of the oxygen did not react with the iron filings?

2. How would your values for the partial pressures of oxygen and nitrogen be affected if you were running this experiment at a higher altitude?

3. Why do the percentages of CO_2 differ in the atmosphere and in expired (exhaled) air?

4. What is the total pressure in mm Hg of a sample of gas that contains 45 mm Hg O_2, 1.20 atm N_2, and 825 mm Hg He?

5. A mixture of gases has a total pressure of 1650 mm Hg. The gases in the mixture are helium at a pressure of 215 mm Hg, nitrogen at a pressure of 0.28 atm, and oxygen. What is the partial pressure of the oxygen in atm?

EXPERIMENT 20
CHARACTERISTICS OF SOLUTIONS

GOALS

1. Determine the polarity of a solute by its solubility in polar and nonpolar solvents.
2. Determine the effect of particle size, stirring and temperature upon the rate of solution formation.
3. Identify an unsaturated and a saturated solution.
4. Graph the solubility curve for KNO_3.

MATERIALS NEEDED

test tubes
test tube rack
small beakers
thermometer
ice

$KMnO_4$ (s)
sucrose
I_2 (s)
NaCl (s), coarse and fine
cyclohexane
vegetable oil
KNO_3 (s)

CONCEPTS TO REVIEW

solute and solvent
formation of solutions
polar and nonpolar solutes
saturated solution
solubility
rate of solution

BACKGROUND DISCUSSION

A solution is composed of a substance called the *solute* dissolved in another substance called the *solvent*. The solution forms because the attractive forces between the solute and the solvent are similar. Thus, a polar (or ionic) solute dissolves in a polar solvent, while a nonpolar solute requires a nonpolar solvent. A substance that is a solid, a liquid, or a gas may be a solute or a solvent. Carbonated beverages are solutions of CO_2 gas in water. Air is a mixture of oxygen and nitrogen gases. However, most of the solutions we work with are liquid. Water, the solvent of the oceans and the body fluids, is known as the *universal solvent*.

Some solutions form quickly and others form slowly. The rate of solution depends upon the size of particle, the amount of shaking or stirring that occurs, and the temperature.

EXPERIMENT 20

A solution may be *dilute* and contain a small amount of solute, or it may be *concentrated* and contain a large amount. When a solution holds the most it can at a certain temperature, it is *saturated*. When more solute is added, it will not dissolve and will appear as a solid in the container. Generally, the amount of solute that dissolves can be increased by heating the solution. Most solutes are more *soluble* at higher temperatures.

LABORATORY ACTIVITIES

Laboratory goggles must be worn!

A. POLARITY OF SOLUTES AND SOLVENTS

This may be a demonstration in the laboratory.

A.1 Set up four test tubes in the test tube racks. Fill each about 1/3 -1/2 full with water, a polar solvent.

Solubility of solutes. Using tweezers or a small spatula, add a few crystals of a different solute (or a few drops), $KMnO_4$, I_2, sucrose, and vegetable oil.

IODINE CAN BURN THE SKIN. HANDLE CAUTIOUSLY!

Stir each mixture with a glass stirring rod. From the appearance of the resulting mixture, indicate whether the solute dissolves in the water, a polar solvent.

A.2 Repeat the procedure using four dry, clean test tubes. Place 3-4 mL of cyclohexane, a nonpolar solvent, in each.

CYCLOHEXANE IS FLAMMABLE: DO NOT PROCEED IF LABORATORY BURNERS ARE IN USE.

Add the same solutes to the cyclohexane, a nonpolar solvent, one to each test tube, and record your observations.

A.3 Describe the polarity of each solute from its behavior in the polar solvent water, and the nonpolar solvent cyclohexane. If a solute dissolves in a polar solvent, the solute is a polar solute; if it dissolves in a nonpolar solvent, it is a nonpolar solute.

B. FACTORS AFFECTING THE RATE OF SOLUTION
You may wish to work in pairs on the following experiments.

B.1 *Effect of Particle Size.* Add some finely ground NaCl crystals in a test tube to a depth of about 1 cm. To another test tube, add some coarse NaCl crystals to the same depth. Add 10 mL water to each. Begin stirring *both* samples with glass stirring rods. Record the time it takes to dissolve the salt in each of the test tubes.

B.2 ***Effect of Stirring.*** Fill two test tubes with small, but equal amounts of finely ground NaCl (about 1 cm depth). Add 10 mL of tap water to each. Stir only one of the samples. State which sample dissolves first.

B.3 ***Effect of Temperature.*** Fill two beakers of the same size about 2/3 full mL of water. Heat the water in one of the beakers to about 80°C. Obtain a few crystals (6 or 8) of $KMnO_4$. Place both beakers (cold and hot) where they will not be disturbed. Add 3-4 $KMnO_4$ crystals to each beaker. Describe the appearance of the two solutions at 5, 10, and 20 minutes.

C. SOLUBILITY OF KNO_3 AT DIFFERENT TEMPERATURES

C.1 Obtain a piece of weighing paper or a small container and weigh carefully.

C.2 Place between 2.50-3.50 g of KNO_3 on the paper or in the container. Weigh to the nearest 0.1 g. Record. Calculate the mass of KNO_3.

C.3 Place 10.0 mL of water in a large test tube. Add the KNO_3. If all of the KNO_3 sample dissolves, place the test in an ice-water mixture. Stir the mixture gently with a thermometer noting the temperature at which the first crystals of solid KNO_3 appear. This is the temperature at which this solution became saturated. Record the temperature.

If the added solid does not completely dissolve at room temperature, prepare a water bath (beaker of water). With the test tube in the water, heat gently until all of the solid dissolves. Stir. Remove the test tube and allow the solution to cool while you stir the solution gently with the thermometer. Record the temperature at which the first crystals appear.

Weigh out four more samples of KNO_3, each weighing between 2.50 g and 3.50 g. Add one of the weighed-out samples to the solution already in the large test tube. Warm until the solid completely dissolves. Then remove and cool while stirring. If necessary, use an ice-water mixture. Note the temperature at which the crystals first begin to form. This is the temperature of saturation for this new, more concentrated solution. Add another sample of KNO_3 and heat and cool as before. Record the temperature for the appearance of the first crystals for each sample.

Calculations of Solubility

C.4 Calculate the *total mass of KNO_3* present in each solution. Add the new quantity to what is already in the test tube.

C.5 Express the solubility of each solution which is the maximum g of solute in 100 g of water. Since we used 10.0 mL (= 10.0 g) of water, the mass of solute in each must be multiplied by 10.

$$\frac{g\ KNO_3}{10.0\ g\ water} \times \frac{10}{10} = \frac{g\ KNO_3}{100\ g\ water} = \text{solubility}$$

C.6 Use the data from above to prepare a graph of the solubility curve for KNO_3. Plot the solubility in g KNO_3/100 g water on the vertical axis against the temperature (0-100°C) on the horizontal axis.

NAME_____ SECTION_____ DATE_____

EXPERIMENT 20
CHARACTERISTICS OF SOLUTIONS
LABORATORY REPORT

A. POLARITY OF SOLUTES AND SOLVENTS

Solute	Soluble in water	cyclohexane	Type of Solute (polar/nonpolar)
$KMnO_4$	_____	_____	_____
I_2	_____	_____	_____
sucrose	_____	_____	_____
vegetable oil	_____	_____	_____

State the general solubility rule concerning the polarities of a solute and solvent.

_____

_____

B. FACTORS AFFECTING THE RATE OF SOLUTION

B.1 Effect of Particle Size

Type of NaCl	Time needed to completely dissolve
fine	_____ min
coarse	_____ min

Which sample dissolved faster? _____

Explain how particle size affects the rate of solution.

_____

_____

177

EXPERIMENT 20

B.2 Effect of Stirring

Sample that dissolved first _____
      (stirred or unstirred)

Explain the effect of stirring upon the rate of solution.

_____

_____

B.3 Effect of Temperature

Time	Observations	
	Hot	Cold
5 min	_____	_____
10 min	_____	_____
15 min	_____	_____

What is the effect of temperature upon the rate of solution?

_____

_____

C. SOLUBILITY OF KNO_3 AT DIFFERENT TEMPERATURES

Mass of container	Mass of container + KNO_3	Mass of KNO_3 added	Temperature (°C)	Mass of KNO_3 in solution	Solubility $\frac{g\ KNO_3}{100\ g\ H_2O}$
_____ g	_____ g	_____ g	_____ °C	_____ g	_____
_____ g	_____ g	_____ g	_____ °C	_____ g	_____
_____ g	_____ g	_____ g	_____ °C	_____ g	_____
_____ g	_____ g	_____ g	_____ °C	_____ g	_____
_____ g	_____ g	_____ g	_____ °C	_____ g	_____

NAME_____ SECTION_____ DATE_____

C.6 *Graphing the solubility of KNO_3 versus temperature (°C)*

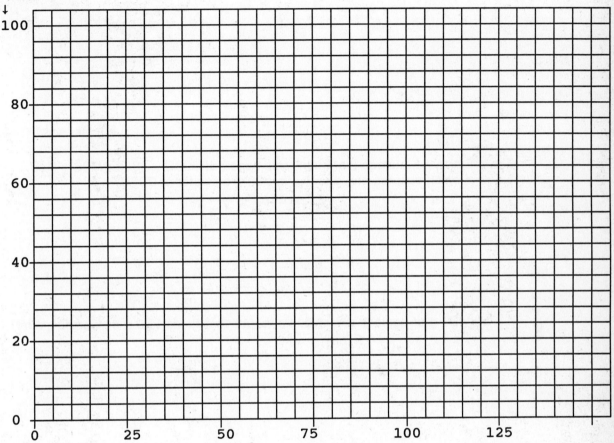

1. According to your graph, how does the solubility of KNO_3 change as the temperature increases?

2. On your solubility curve, what is the change in solubility of KNO_3 from 30°C to 60°C?

3. At what temperature does KNO_3 have a solubility of 50 g/100 g H_2O?

EXPERIMENT 20

QUESTIONS AND PROBLEMS

1. How does an unsaturated solution differ from a saturated solution?

2. The solubility of sucrose (common table sugar) at 70°C is 320 g/100 g H_2O.
   a. How much sucrose can dissolve in 200 g of water at 70°C?

   b. Will 400 g of sucrose dissolve in a teapot that holds 200 g of water? Why?

3. If the solubility of sucrose at 0°C is 180 g/100 g H_2O, will 400 g of sucrose dissolve in a pitcher of 200 g of iced tea at 0°C? If not, how many grams will?

EXPERIMENT 21
CONCENTRATIONS OF SOLUTIONS

GOALS

1. Calculate the percent water in a fruit or vegetable.
2. Measure the volume and mass of a salt solution.
3. Evaporate the solution to dryness and determine the mass of the solute.
4. Calculate the weight/weight percent and weight/volume percent concentrations for the solution.
5. Calculate the molar concentration of the solution.

MATERIALS NEEDED

A. Water in food
fruit or vegetable
knife
drying oven
watch dish

B. Concentrations
ringstand and iron ring
wire screen
evaporating dish
NaCl solution
400 mL beaker
10 mL graduated cylinder, or 10-mL pipet
Bunsen burner

CONCEPTS TO REVIEW

water as a solvent
formation of solutions
percent and molar concentrations

BACKGROUND DISCUSSION

Many of the foods we eat are composed of large quantities of water as are the cells of our bodies which are about 60% water. We obtain water for the body primarily by drinking fluids and eating food with a high content of water. We can calculate the amount of water in a food by dehydrating a piece of food and measuring the amount of water lost.

EXPERIMENT 21

The concentration of a solution is calculated from the amount of solute present in a certain amount of solution. The concentrations may be expressed using different units for amount of solute and solution. A **weight percent** concentration expresses the grams of solute in the grams of solution. The **weight/volume percent** concentration of a solution states the grams of solute present in the milliliters of the solution.

$$\text{weight percent} = \frac{\text{grams of solute}}{\text{grams of solution}} \times 100$$

$$\text{weight/volume percent} = \frac{\text{grams of solute}}{\text{milliliters of solution}} \times 100$$

A *molar (M) concentration* describes a solution in moles of the solute per liter of solution.

$$\text{molarity (M)} = \frac{\text{moles of solute}}{\text{liters of solution}}$$

In this experiment, you will measure out a 10.0 mL volume of a sodium chloride solution. The mass of the solution will be determined by weighing the solution in a pre-weighed evaporating dish. After the sample is evaporated to dryness, it is again weighed in the evaporating dish. From this data, the mass of the salt (solute) is obtained.

Using the mass of the solute and the mass of the solution, the weight percent can be calculated. From the mass of the solute and the volume (mL) of the solution, the weight/volume percent can be calculated. The molarity of the solution is obtained by converting the mass of the solute to moles, and expressing the volume in liters (L).

LABORATORY ACTIVITIES

Put on your safety glasses before you begin this lab!

A. PERCENT WATER IN A FOOD

A.1 Weigh a watch glass to the nearest 0.01 g.

A.2 Cut several *thin* slices of one fruit or vegetable. Record the type of food.

A.3 Arrange the food slices on the watch glass as far apart as possible. Weigh the watch glass and food slices. Record.

A.4 Place the watch glass and food in a drying oven at 100-110°C for 1 hr. Carefully remove the watch glass (it will be hot!) and dried (dehydrated) food. Let it cool to room temperature. Weigh the watch glass and dried food and record the total mass to the nearest 0.01 g.

Calculations

A.5 Calculate the mass (to 0.01 g) of the food slices (A.3 - A.1).

A.6 Calculate the mass of the dried food slices (A.4 - A.1).

A.7 Calculate the amount of water lost through dehydration (A.5 - A.6).

A.8 Calculate the percentage of water in the food.

$$\% \text{ water} = \frac{\text{mass of water lost (A.7)}}{\text{mass of food slices (A.5)}} \times 100\%$$

A.9 List the values for other foods tests. Look up handbook values for the percent water in the foods that were tested.

B. CONCENTRATION OF A SODIUM CHLORIDE SOLUTION

B.1 Weigh a dry evaporating dishes to 0.01 g. Record the mass.

B.2 Using a 10.0 mL graduated cylinder, or a 10.0 mL pipet, measure out a 10.0 mL sample of the NaCl solution provided. Record.

B.3 Weigh the evaporating dish and the NaCl solution to 0.01 g. Record.

B.4 Fill a 400 mL beaker about one-half full of water. Set on a wire screen and place the evaporating dish on top of the beaker. Heat the water in the beaker to boiling. See Figure 21.1. You may need to replenish the hot water bath as you proceed.

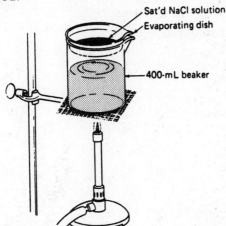

Figure 21.1 *Apparatus for evaporation of salt solution.*

When the NaCl appears to be dry, **carefully** remove the evaporating dish, dry the bottom, and set the dish directly on the wire screen. Heat gently with a low flame to completely dry the salt. Cool for 10 minutes. Weigh the evaporating dish and the dry NaCl.

EXPERIMENT 21

Calculations

B.5   Calculate the mass of the NaCl after drying (B.4 - B.1).

B.6   Calculate the mass of the solution (B.3 - B.1).

B.7   Calculate the weight/weight percent concentration.

$$\text{weight/weight percent} = \frac{\text{dry mass of NaCl (B.5)}}{\text{mass (g) of solution (B.6)}} \times 100$$

B.8   Calculate the weight/volume percent concentration.

$$\text{weight/volume percent} = \frac{\text{dry mass of NaCl}}{\text{volume of solution (10.0 mL)}} \times 100$$

B.9   Calculate the moles of NaCl (58.5 g/mole).

$$\text{g NaCl} \times \frac{1 \text{ mole NaCl}}{58.5 \text{ g NaCl}} = \text{moles NaCl}$$

B.10   Convert the volume of the solution to liters.

$$\text{mL NaCl solution} \times \frac{1 \text{ L}}{1000 \text{ mL}} = \text{L solution}$$

B.11   Calculate the molarity of the NaCl solution.

$$\text{molarity (M)} = \frac{\text{moles NaCl}}{\text{L of solution}}$$

NAME_____ SECTION_____ DATE_____

EXPERIMENT 21
CONCENTRATIONS OF SOLUTIONS
LABORATORY REPORT

A. PERCENT WATER IN A FOOD

A.1　Type of food　　　　　　　　　　_____

A.2　Mass of watch glass and food　　_____ g

A.3　Mass of watch glass　　　　　　 _____ g

A.4　Mass of watch glass and dry food _____ g

Calculations:

A.5　Mass of food　　　　　　　　　　_____ g

A.6　Mass of dry food　　　　　　　　_____ g

A.7　Water loss by dehydration　　　　_____ g

A.8　Percent water in food　　　　　　_____ g

A.9　Food　　　　　% Water (Experimental)　　　　% Water (Handbook)

　　_____　　_____　　　　_____ g

　　_____　　_____　　　　_____ g

　　_____　　_____　　　　_____ g

　　_____　　_____　　　　_____ g

　　_____　　_____　　　　_____ g

EXPERIMENT 21

B. CONCENTRATION OF A SODIUM CHLORIDE SOLUTION

B.1 Mass of evaporating dish _____ g

B.2 Volume of NaCl solution _____ mL

B.3 Mass of dish + NaCl solution _____ g

B.4 Mass of dish and dried NaCl _____ g

B.5 Mass of the dried NaCl _____ g
Show calculations:

B.6 Mass of NaCl solution _____ g

B.7 Weight/weight percent _____ %
Show calculations:

B.8 Weight/volume percent _____ %
Show calculations:

B.9 Moles of NaCl _____ moles
Show calculations:

B.10 Volume (L) of sample _____ L

B.11 Molarity _____ M
Show calculations:

QUESTIONS AND PROBLEMS

1. 15.0 mL of a NaCl solution that has a mass of 15.78 g is placed in an evaporating dish and evaporated to dryness. The residue has a mass of 3.26 g. Calculate the following concentrations for the NaCl solution.

    a. weight percent

    b. weight/volume percent

    c. molarity

2. A 3.0 % (w/v) KI has a volume of 25.0 mL. How many grams KI are in the sample?

3. How many g of a 25% (w/w) NaCl solution contain 150 g NaCl?

4. What is the molarity of a solution that contains 80.0g of NaOH (molar mass = 40.0 g/mol) dissolved in 500 mL of solution?

EXPERIMENT 22
SOLUTIONS, COLLOIDS, AND SUSPENSIONS

GOALS

1. Observe the transport systems in solutions.
2. Determine the ability of suspension particles, colloids, and solution particles to pass through filters and/or membranes.
3. Distinguish between osmosis and dialysis.

MATERIALS NEEDED

A.
beaker
test tubes
test tube rack
10% starch solution
10% NaCl
10% glucose
charcoal
funnel
filter paper (Whatman #1)
0.1 M $AgNO_3$
Benedict's reagent
iodine reagent

B. 10% starch
charcoal
250-mL beaker
50-mL graduated cylinder
iodine reagent
funnel, filter paper

C.
20 cm dialysis bag
distilled water
10% glucose
10% starch
10% NaCl
250 mL beaker
test tubes (3)
funnel
boiling water bath
Benedict's reagent
0.1 M $AgNO_3$
iodine reagent

CONCEPTS TO REVIEW

solutions
colloids
suspensions
osmosis
dialysis

EXPERIMENT 22

BACKGROUND DISCUSSION

Materials move in and out of cells by (1) *filtration* produced by gravity; (2) *osmosis* whereby water moves across a semipermeable membrane; and (3) *dialysis* whereby solution particles and water move across a semipermeable membrane when concentration gradients exist.

In this experiment, you will first determine a test that can be used to identify the presence of the substances that will be in the solutions. In part B of the experiment, you will mix some charcoal with a starch solution. You will determine how the charcoal, a suspension, separates from the starch, a colloid.

In Experiment C, *dialysis* will be demonstrated by the movement of true-solution particles through a semipermeable membrane while colloidal particles are retained. Many of the membranes in the body are dialyzing membranes. For example, the intestinal tract consists of a semipermeable membrane that allows the simple, solution particles from digestion to pass into the blood and lymph. Larger, incompletely digested food particles that are colloidal size or larger remain within the intestinal wall. Dialyzing membranes are also used in hemodialysis to separate waste particles, particularly urea, out of the blood.

LABORATORY ACTIVITIES

WEAR SAFETY GOGGLES

A. IDENTIFICATION TESTS

In this experiment, you will need to carry out tests that indicate the presence or absence of certain substances. The tests you perform in this section, and the results you observe, will be used as a reference for identifying those substances in other parts of this experiment.

Preparation of reagent containers. Small amounts of the reagents for the tests may be placed in small beakers or vials for use throughout this experiment. *Dropper bottles of these reagents may be available in the laboratory for your use.* If not, place 3-4 mL of 0.1 M $AgNO_3$ in a vial. Label. Place 25-30 mL of Benedict's reagent in another beaker. In another vial or small container, such as a shell vial, place 3-4 mL of iodine reagent. Keep these reagent containers at your desk for the duration of the experiment.

 Be careful in using $AgNO_3$ and the iodine reagent. They both stain.

In a small beaker, prepare a mixture of 5 mL of 10% glucose, 5 mL of 10% NaCl, and 5 mL of a 10% starch solution. After stirring the mixture, pour 5 mL of the mixture into each of three test tubes. To three other test tubes, add 5 mL of distilled water.

A.1 First test tube: Test for Cl⁻ in the mixture by adding 5 drops of 0.1 M $AgNO_3$. Also add 5 drops of 0.1 M $AgNO_3$ to a test tube containing distilled water.

Second test tube: Test for starch by adding 3-4 drops of iodine reagent. Also add 3-4 drops of iodine reagent to a test tube containing distilled water.

Third test tube: Test for glucose by adding 5 mL of Benedict's reagent to the mixture, and to a test tube containing distilled water. Heat the test tubes in a boiling water bath for 5 min.

After you have completed each of the above tests, describe the test results in the test tube containing the mixture, and in the test tubes containing distilled water. Record the colors of the solution, new colors, and whether any precipitates formed. A *positive test* has occurred when a change has occurred in the original properties of the reagents. If there is no change in the reagents, it is a *negative test*.

B. FILTRATION

B.1 In a small beaker, prepare a mixture of 5 mL of the 10% starch solution and a small amount of charcoal, and 10 mL water. Place a piece of filter paper in a funnel and set the funnel in a test tube. Pour the mixture into the filter paper in the funnel. Collect the liquid(filtrate) that passes through the paper into the test tube.

Observe the filter paper and describe what you see.

B.2 Test the liquid in the test tube (filtrate) for starch by adding a few drops of iodine reagent. Compare the result of the iodine test to the results in part A. If you observe the same test results for starch, then starch is *present* in the liquid that filtered through the paper.

B.3 From the test results, determine whether starch and/or charcoal were trapped in the filter, or passed through. Identify each as a colloid or a suspensions.

C. DIALYSIS

Prepare a mixture of 5 mL of 10% glucose, 5 mL of 10% NaCl, and 5 mL of a 10% starch solution. Obtain a dialysis bag that has been soaking in distilled water. Tie a knot in one end. Place a funnel in the open end and pour in the remaining mixture. Tie a firm knot in the open end to close the dialysis bag. Rinse the dialysis bag with distilled water. Place the dialysis bag in a 250 mL beaker containing 100 mL distilled water. See Figure 22.1.

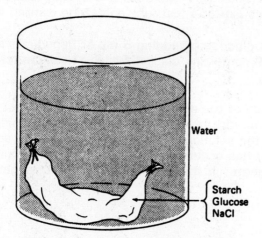

Figure 22.1 Dialysis bag placed in distilled water.

C.1 **Immediately** pour off 15 mL of the distilled water surrounding the bag for the first group of tests. Divide the 15 mL into three test tubes (5 mL each). Repeat the tests from part A.

> Test the water in the first test tube for Cl⁻ by adding $AgNO_3$.
> Test the water in the second test tube for starch by adding iodine reagent.
> Test the water in the third test tube for glucose.

Repeat the above test on new 15 mL samples of water from *outside* the dialysis bag at 20 min, 40 min, and 60 min. Separate into three test tubes and test for Cl-, starch, and glucose again. Each time record your test results. If one of these substances gives a positive test, mark it as present(+); if there are no changes, mark the substance as absent(-).

After the final test, break the bag open, and test 15 mL of the contents. From your results, determine which substance(s) dialyzed through the membrane, and which substance(s) were retained.

NAME _____ SECTION _____ DATE _____

EXPERIMENT 22
SOLUTIONS COLLOIDS AND SUSPENSIONS
LABORATORY REPORT

A. IDENTIFICATION TESTS

A.1   Test results: Mixture   Distilled water
                                   (Positive test)   (Negative test)

      Test for Cl^- _____   _____

      Test for starch   _____   _____

      Test for glucose   _____   _____

B. FILTRATION

B.1   Appearance of filter paper   _____

      Substance present   _____

B.2   Appearance of filtrate
      after adding iodine reagent   _____

      Substance present   _____

B.3   Indicate the type of mixtures represented by charcoal and starch

              Mixture   Type (suspension or colloid)

              charcoal   _____

              starch   _____

C. DIALYSIS

C.1 Results of Testing Dialysate

Time	Cl⁻	starch	glucose
0 min			
20 min			
40 min			
60 min			

	Cl⁻	starch	glucose
Contents of dialysis bag			

1. Which substance(s) were found in the water *outside* the dialysis bag?

2. How were those substance(s) able to get through the dialysis bag?

3. What substance(s) were retained inside the dialysis bag? Why?

NAME_____ SECTION_____ DATE_____

QUESTIONS AND PROBLEMS

1. Osmosis occurs when water moves through the walls of red blood cells. When a red blood cell is placed in a *isotonic* solution where the osmotic pressure is the same as that inside the red blood cell, the flow of water is equal into and out of the cell. Both 0.9% NaCl (saline) and 5% glucose solutions are considered isotonic to the cells of the body. When a cell is placed in a *hypotonic* or a *hypertonic* solution, the flow of water in and out of the cell is no longer the same, and the cell volume is altered. State whether the following concentrations of various solutions are isotonic, hypotonic, or hypertonic:

   a. H_2O _____

   b. 0.9% NaCl _____

   c. 10% glucose _____

   d. 3% NaCl _____

   e. 0.2% NaCl _____

2. A red blood cell in a hypertonic solution will shrink in volume as it undergoes *crenation*. In a hypotonic solution, a red blood cell will swell and possibly burst as it undergoes hemolysis. Predict the effect on a red blood cell (crenation, hemolysis, or none) for a red blood cell in the following solutions:

   a. 2% NaCl _____

   b. H_2O _____

   c. 5% glucose _____

   d. 1% glucose _____

   e. 10% glucose _____

3. Why are only isotonic solutions used as parenteral solutions? A parenteral solution is any solution not given orally.

EXPERIMENT 23
ELECTROLYTES AND INSOLUBLE SALTS

GOALS

1. Compare the conductivities of strong, weak and nonelectrolytes.
2. List the electrolytes and their concentrations (mequiv/L) in intravenous solutions.
3. Test a variety of water samples for water hardness.
4. Use water treatment techniques to purify water.

MATERIALS NEEDED

A. *Electrolytes (classroom demonstration)*
conductivity apparatus
mineral water
0.1 M NaOH
0.1 M HCl
0.1 M HAc
0.1 M NH_4OH
0.1 M NaCl
1 M sucrose
1 M glucose
ethanol
soft drink

B. *Electrolytes in body fluids*
IV solutions

C. *Testing the hardness of water*
test tubes and test tube rack
tincture of green soap
water samples: distilled, hard, soft, tap, sea water, mineral water, fish tank, pool, hot tub, lake, river

D. *Purification of water*
muddy water
1% $Al_2(SO_4)_3$
1% Na_2SO_4
1% $FeCl_3$
1% NaCl

CONCEPTS TO REVIEW

solutions
electrolytes and nonelectrolytes
milliequivalents

BACKGROUND DISCUSSION

Electrolytes are substances that produce ions in water. When ions are present in an aqueous solution, the light bulb of a conductivity apparatus will glow, because the ions complete the electrical circuit. A nonelectrolyte produces only molecular substances which do not carry current in an aqueous solution. The light bulb in the conductivity apparatus does not glow. Weak electrolytes produce a few ions: the light bulb will glow weakly. We can identify substances in aqueous solutions as electrolytes, weak electrolytes, or nonelectrolytes on the basis of our observations of the conductivity apparatus.

EXPERIMENT 23

The cells of the body are bathed both inside and outside by fluids that contain specific, differing amounts of electrolytes. The electrolytes play an important role in the maintenance of the activities in the cell. Electrolytes that are positive ions are called *cations*, negative ions are *anions*. Inside the cell, the major cation is potassium, K^+, and the major anion is bicarbonate, HCO_3^-. The fluid within the cell is called the *intracellular fluid*. Outside the cell, in the *extracellular fluid*, the major cation is sodium Na^+, and the major anion is chloride, Cl^-. Other electrolytes are found in smaller quantities in both intracellular and extracellular fluids.

When there is a loss of fluid from the body or an imbalance of electrolytes, a parenteral solution (one given by means other than oral) may be administered. The type of parenteral solution reflects the needs of the cells in the body as determined by laboratory tests of the body fluids. When giving parenteral solutions such as intravenous solutions, keep in mind the effect of the electrolytes in maintaining and regulating fluid balance, muscle tone, and acid-base balance in the body. See Table 23.1. If there is an imbalance, there may also be a shift of water between plasma and tissues or a loss of water accompanied by a shift or loss of essential electrolytes.

Table 23.1   Effects of Electrolytes Imbalance

Electrolyte	Low levels	High levels
Na^+, Cl^-	weakness, headache, diarrhea, cramping	edema
K^+	apathy, cardiac changes	cardiac arrest, numbness
Ca^{2+}	numbness in extremities, tetany	deep bone pain
HCO_3^-	acidosis (low pH)	alkalosis (high pH)
Mg^{2+}	convulsions, disorientation	dehydration, coma

Water that contains the ions Ca^{2+}, Mg^{2+}, and Fe^{3+} is called **hard water**. When hard water reacts with soap, the ions in the hard water and some of the soap molecules form insoluble salts called **soap scum**. The soap molecules tied up in the scum are not free to perform their cleaning function. More soap must be added to remove the ions and allow sufficient sudsing and cleaning.

The reaction of a soap solution with the ions in hard water can be used to compare hardness of water samples. In some hospitals, a soap solution may be used to test the water used for dialysis treatments, which must be free of these ions. In water treatment stations, certain chemicals added to water will cause the formation of insoluble substances that sink to the bottom of the tank. The water on top is purified and can be drawn off for use.

LABORATORY ACTIVITIES

A. ELECTROLYTES AND CONDUCTIVITY

 THIS PART OF THE EXPERIMENT WILL BE A DEMONSTRATION BY YOUR INSTRUCTOR. THE BARE ELECTRODES ARE A HAZARD! DO NOT TOUCH THE ELECTRODES WHEN THE APPARATUS IS PLUGGED IN. YOUR SKIN WILL CONDUCT AN ELECTRIC CURRENT AND CAUSE A SHOCK.

A.1 Select one of the solutions to test. Place the electrodes in the solution and observe the light bulb. Identify the solution as a strong electrolyte, a weak electrolyte, or a nonelectrolyte. See Figure 23.1. Select another substance and record the intensity of the light bulb for it. Repeat the test for electrolytes with each solution.

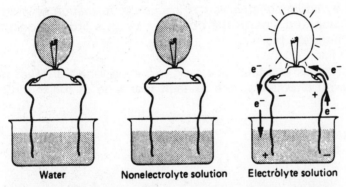

Figure 23.1 Light bulb apparatus for conductivity determination.

A.2 On the laboratory record, indicate the kinds of particles that must be present in each solution. Strong electrolytes contain only ions in solution, weak electrolytes contain molecules and a few ions, while nonelectrolytes are dissolved as molecules.

B. ELECTROLYTES IN BODY FLUIDS

B.1 In the laboratory, you should find a display of bottles containing intravenous (IV) solutions. Record the type of solutions in three of these.

B.2 Record the types of electrolytes present in each solution and give their concentrations in mequiv/L. These are listed on the label by their symbols. For example, Na 47 Cl 47 means that there are 47 mequiv of sodium ion and 47 mequiv chloride ion per liter of the solution.

$$\text{Na 47 Cl 47} = \text{Na}^+, 47 \text{ mequiv/L} \quad \text{Cl}^-, 47 \text{ mequiv/L}$$

B.3 Calculate the total number of meq of cations and record; calculate the total number of meq of anions and record.

B.4 Calculate the overall sum of the positive and negative charges and record.

C. TESTING THE HARDNESS OF WATER

Obtain 100-150 mL of a soap solution in a beaker. Set up a buret, and fill with the soap solution. Your instructor will demonstrate this procedure. Place 50 mL of the water sample you are going to test (begin with distilled water) in a 250 mL flask. Add 1 mL of the soap solution from the buret. Stopper the flask and shake for 10 sec. With distilled water, you should see a thick layer of suds. If you don't, add another milliliter of soap solution and shake for 10 sec again. The suds that form in the distilled water sample will serve as your reference sample. Save for comparison. Reshake if necessary.

Test several of the other water samples. If no suds form, continue to add soap solution until you have softened the water sample, and it forms suds like the distilled water sample. Use an assortment of water samples available in the lab or from your home, pool or well. Record the number of mL required to soften each water sample.

D. PURIFICATION OF WATER

Water can be purified by adding chemicals that cause the formation of large particles that act like a suspension and sink to the bottom. Such a process occurs in the settling tanks at a water filtration facility.

D.1 Set up five test tubes in a test tube rack. To each, add 10 mL of muddy water. Add the following chemicals, one to each test tube of the muddy water; label each.

(1) 5 mL water
(2) 5 mL 1% NaCl
(3) 5 mL 1% $Al_2(SO_4)_3$
(4) 5 mL 1% Na_2SO_4
(5) 5 mL 1% $FeCl_3$

Stir each test tube thoroughly. Look for a separation of a precipitate and a clarification of the upper portion of water. Record your observations at 10 min, 30 min and 60 min intervals.

NS no settling, still muddy
BS beginning to settle, still murky
SS some settling, cloudy
MS mostly settled, slightly cloudy
SC settled, clear

NAME_____ SECTION_____ DATE_____

EXPERIMENT 23
ELECTROLYTES AND INSOLUBLE SALTS
LABORATORY REPORT

A. ELECTROLYTES AND CONDUCTIVITY

Substance	A.1 type of electrolyte (strong, weak, non)	A.2 Type of particles in solution (ions, molecules, both)
tap water		
0.1 M NaCl		
0.1 M NaOH		
0.1 M acetic acid (HAc)		
1 M sucrose		
0.1 M HCl		
0.1 M NaOH		
0.1 M NH_4OH		
C_2H_5OH ethanol		
1 M glucose		
mineral water or soft drink		

EXPERIMENT 23

B. ELECTROLYTES IN BODY FLUIDS

B.1 Type of IV solution

_____ (1) _____ (2) _____ (3)

B.2 Listing of electrolytes (mequiv/L)

cations(+)	anions(-)	cations(+)	anions(-)	cations(+)	anions(-)
_____	_____	_____	_____	_____	_____
_____	_____	_____	_____	_____	_____
_____	_____	_____	_____	_____	_____
_____	_____	_____	_____	_____	_____
_____	_____	_____	_____	_____	_____

B.3 Total charge

_____ _____   _____ _____   _____ _____

B.4 Overall charge

_____   _____   _____

C. TESTING THE HARDNESS OF WATER

Type of water	mL of soap	Type of water	mL of soap
distilled water	_____	mineral water	_____
tap water	_____	sea water	_____
_____	_____	_____	_____
_____	_____	_____	_____

Which water sample was the softest? Why?

Which water sample was the hardest? Why?

NAME_____ SECTION_____ DATE_____

D. PURIFICATION OF WATER

Settling agent	Appearance after time intervals		
	10 min	30 min	60 min
H_2O			
NaCl			
$Al_2(SO_4)_3$			
Na_2SO_4			
$FeCl_3$			

How does the presence of ions affect the rate of settling compared to the rate of settling with water?

Which chemical produced the most rapid settling?

EXPERIMENT 23

QUESTIONS AND PROBLEMS

1. Identify the types of particles in each of the following solutions. State whether each of the following equations represents a weak electrolyte, a strong electrolyte, or a nonelectrolyte.

a. $XY_2 \underset{\longleftarrow}{\overset{H_2O}{\longrightarrow}} X^{2+}(aq) + 2Y^-(aq)$

b. $HX \xrightarrow{H_2O} H^+(aq) + X^-(aq)$

c. $XYZ(s) \longrightarrow XYZ(aq)$

d. $YOH(s) \longrightarrow Y^+(aq) + OH^-(aq)$

2. In part A, you tested solutions of different substances for electrolytes. Write an equation for the solution of the following in water:

HCl _____

NaOH _____

glucose, $C_6H_{12}O_6$ _____

acetic acid
$HC_2H_3O_2$ or HAc _____

EXPERIMENT 24
TESTING FOR CATIONS AND ANIONS

GOALS

1. Determine the presence of a cation (positive ion) or anion (negative ion) by a a chemical reaction.
2. Determine the presence of some cations and anions in a solution of an unknown salt.
3. Summarize the solubility rules for insoluble salts.

MATERIALS NEEDED

test tubes
test tube rack
small beakers
ringstand and iron ring
flame-test wire
red litmus paper
0.1 M KCl
0.1 M NaCl
0.1 M $CaCl_2$
0.1 M NH_4Cl
0.1 M $FeCl_3$
0.1 M $(NH_4)_2C_2O_4$
0.1 M KSCN
6 M HCl
6 M NaOH

0.1 M Na_3PO_4
0.1 M Na_2SO_4
0.1 M Na_2CO_3
0.1 M $AgNO_3$
6 M HNO_3
0.1 M $(NH_4)_2MoO_4$
0.1 M $BaCl_2$

CONCEPTS TO REVIEW
ions
chemical change
solubility rules

BACKGROUND DISCUSSION

Many solutions such as the milk and the juices you drink contain an assortment of anions and cations. Many of these ions have distinctive chemical reactions. Therefore, we can perform tests in which observable changes can be attributed to the presence of a particular ion. In this experiment, you will observe chemical changes such as the formation of a solid (precipitate), a change in color, the formation of bubbles(gas), or the color in a flame test. Your observations will be the key to identifying those same ions when you test unknown solutions.

The cations you will be testing are the following:

cations	solution
sodium (Na^+)	NaCl
potassium (K^+)	KCl
calcium (Ca^{2+})	$CaCl_2$
iron (Fe^{3+})	$FeCl_3$
ammonium (NH_4^+)	NH_4Cl

The presence of sodium, potassium, and calcium ions can be determined by their distinctive colors in flame tests. Calcium can be confirmed by the addition of $(NH_4)_2C_2O_4$ to give a white precipitate. Iron is detected by the distinctive color it forms with potassium thiocyanate, KSCN. When the ammonium ion is converted to ammonia, it gives a distinctive odor, and turns red litmus blue.

The anions tested will be:

anions	solution
chloride (Cl^-)	NaCl
phosphate (PO_4^{3-})	Na_3PO_4
sulfate (SO_4^{2-})	Na_2SO_4
carbonate (CO_3^{2-})	Na_2CO_3

In the test with $AgNO_3$, most of these ions will form precipitates of their silver salts, AgCl, Ag_3PO_4, and Ag_2CO_3. Ag_2SO_4 is more soluble, and should not form a solid. When nitric acid, HNO_3, is added to each of the precipitates, only AgCl will not dissolve. When barium chloride, $BaCl_2$, is added to each of the anions, the precipitates formed are $BaCO_3$, $BaSO_4$, and $Ba_3(PO_4)_2$. $BaCl_2$ remains soluble. When nitric acid, HNO_3, is added to the precipitates, only $BaSO_4$ does not dissolve. This confirms the presence of sulfate (SO_4^{2-}). The carbonate anion (CO_3^{2-}) can be confirmed by the addition of HNO_3 which produces bubbles of CO_2 which have no odor. The phosphate anion (PO_4^{3-}) is confirmed by the addition of ammonium molybdate to give a yellow precipitate.

As you proceed, you will test solutions of knowns that contain a particular ion. Each student will receive an unknown sample containing a cation and an anion. After you observe reactions of knowns, you will carry out the same tests with your unknown. Therefore, you should expect to see a test that matches the reactions of one of the cations, and another for one of the cations. By identifying the ions in your unknown, you will be able to write the name and formula of your unknown salt.

LABORATORY ACTIVITIES

A: TESTS FOR CATIONS

LABORATORY GOGGLES MUST BE WORN DURING THIS EXPERIMENT!

In many parts of this experiment, you will be using about 2 mL portions of different solutions. Place 2 mL of water in the same size test tube that you will be using. As you obtain solutions for the experiment, take a volume that matches the height of the 2 mL water sample.

TESTING FOR CATIONS AND ANIONS

Obtain eight test tubes, cleaned and rinsed with distilled water. Be sure to label each test tube. Many test tubes have a frosted section for writing on. Otherwise use a marking pencil or a label. In five of the test tubes, place 2 mL portions of the following knowns in each of the test tubes: 0.1 M NaCl, 0.1 M KCl, 0.1 M $CaCl_2$, 0.1 M $FeCl_3$, and 0.1 M NH_4Cl. In three other test tubes, place 2 mL portions of the unknown solution obtained from your instructor.

Before you begin this group of experiments, you may also wish to place about 15 mL of each of the following reagents in small beakers: 6 M HCl, 6 M HNO_3, and 6 M NaOH. Another beaker of distilled water is convenient for rinsing out the droppers.

 Once you have removed a solution from the reagent bottle, it must be discarded even if you don't use all of it. NEVER RETURN CHEMICALS TO THEIR STORAGE BOTTLES.

Take a beaker to your instructor for a sample of your unknown. Record the sample number. If the unknown is a solid, dissolve 1 g of the salt in 50 mL of distilled water. Your sample solution is now ready for testing.

A.1 Test for sodium ion, Na^+

Clean a flame test wire by dipping it in the HCl beaker and placing the loop of the wire in a flame. Place the loop in the NaCl solution, then in the flame. The NaCl solution should give a bright, yellow-orange flame.

A.2 Test for potassium ion, K^+

A flame test is also used to identify the presence of K^+. However, the flame is not as persistent or intense as Na^+. Clean the flame test wire again, and then test the KCl solution in the flame. The color of a potassium flame is a pinkish-lavender and does not last long, so you must look for it immediately upon heating the flame test wire. Record the color you observe for the potassium ion, K^+.

A.3 Test for calcium ion, Ca^{2+}

Calcium ion also gives a flame test. Dip a clean flame test wire in the $CaCl_2$ solution, and place in a flame. The $CaCl_2$ solution should give a reddish-orange color. Record.

To confirm the presence of calcium, add 5-10 drops of 0.1 M $(NH_4)_2C_2O_4$, ammonium oxalate to the $CaCl_2$ solution in the test tube. Place the test tube in a warm water bath for 5 minutes. Calcium is confirmed if there is a *slow* formation of a cloudy, white solid(precipitate) which is $CaC_2O_4(s)$. The equation for the reaction is

$$CaCl_2 + (NH_4)_2C_2O_4 \longrightarrow CaC_2O_4(s) + 2NH_4Cl$$

Since only the calcium and oxalate ions react, the net ionic equation is:

$$Ca^{2+} + C_2O_4^{2-} \longrightarrow \underset{\text{white}}{CaC_2O_4(s)}$$

207

EXPERIMENT 24

A.4 *Testing the unknown for Na^+, K^+ or Ca^{2+}*

At this point, repeat the flame test using the unknown solution. Describe any color you observe. If you get a color that matches the color of one of the flame tests, you can identify your cation as Na^+, K^+, or Ca^{2+}. If you think you have calcium, confirm its presence with the confirmation test.

A.5 *Test for ferric ion, Fe^{3+}*

Add 3-4 drops of potassium thiocyanate (KSCN) to the test tube containing $FeCl_3$, and to a 2 mL portion of the unknown. A deep red color should appear if Fe^{3+} is present. A faint pink color is not a positive test for iron. Record results.

$$Fe^{3+} + SCN^- \rightarrow \underset{\text{deep red}}{FeSCN^{2+}}$$

A.6 *Test for ammonium ion, NH_4^+*

Add 20 drops of 6 M NaOH (reagent beaker) to the test tube containing NH_4Cl. Place a strip of moistened red litmus paper across the top of the test tube and set the test tube in a warm water bath. The ammonia gas $NH_3(g)$ that forms will turn the red litmus paper blue. The odor of ammonia may be noticed if you **carefully** waft the vapors from the test tube towards you. Repeat the ammonium ion test with your unknown. Record results.

$$NH_4^+ + OH^- \rightarrow \underset{\text{ammonia}}{NH_3(g)} + H_2O$$

B. TESTS FOR ANIONS

B.1 *Silver nitrate test*

Obtain five test tubes, cleaned and rinsed with distilled water. Be sure to label each test tube. In four test tubes, place 2 mL portions of the following in each of the test tubes: 0.1 M NaCl, 0.1 M Na_3PO_4 and 0.1 M Na_2SO_4, and 0.1 M Na_2CO_3. In the other test tube, place a 2 mL portion of your unknown solution. To each of the test tubes add 4-5 drops of 0.1 M $AgNO_3$. Record your observations for the known solutions, and for your unknown.

BE CAREFUL. $AgNO_3$ STAINS THE SKIN.

Several of the anions should form insoluble silver salts. Record the formation of each precipitate and its color. For example, a white precipitate of AgCl should appear. Record the results of your unknown. *Keep for the next test. Do not discard!* Some sample equations are given below:

$$Ag^+ + Cl^- \longrightarrow \underset{white}{AgCl(s)}$$

$$3\,Ag^+ + PO_4^{3+} \longrightarrow Ag_3PO_4(s)$$

To each of the test tubes containing a *precipitate*, add 10-15 drops of 6 M HNO_3. Stir well with a glass stirring rod. To make sure that each sample is acidic, use a glass rod to place a drop of the acidified solution on a piece of blue litmus paper. The spot should turn red. If not, add a few more drops of HNO_3. Record which precipitates dissolve in HNO_3 and which do not. Record the results of your unknown.

B.2 Test for phosphate ion, PO_4^{3-} (optional)

Place about 2 mL of 0.1 Na_3PO_4 in a test tube, and 2 mL of your unknown in another test tube. Acidify each with 6 M HNO_3 using blue litmus paper. (See above.) Add 1-2 mL of ammonium molybdate solution to each and warm in a hot water bath. A yellow precipitate confirms the presence of the phosphate ion. Record the results of the known and the unknown.

B.3 Barium chloride test

Obtain 5 test tubes, cleaned and rinsed with distilled water. Be sure to label each test tube. In 4 of the test tubes, place 2 mL portions of the following: 0.1 M NaCl, , 0.1 M Na_3PO_4, 0.1 M Na_2SO_4, and 0.1 M Na_2CO_3. In the other test tube, place 2 mL portion of your unknown solution. To each add about 1 mL (20 drops) of 0.1 M $BaCl_2$. Record your observations of any precipitates and their color. For example, in the SO_4^{2-} solution, a fine, white precipitate of $BaSO_4$ should appear.

$$Ba^{2+} + SO_4^{2-} \longrightarrow \underset{white}{BaSO_4(s)}$$

To each of the test tubes that form a precipitate, add 2-3 mL of 6 M HNO_3 to make the solution acidic. All of the precipitates except $BaSO_4$ should dissolve. $BaSO_4$ remains insoluble. Record your results for the knowns and the unknown.

B.4 Carbonate ion, CO_3^{2-}

Place 4-5 mL of 0.1 M Na_2CO_3 in one test tube, and 4-5 mL of your unknown in another. While **carefully observing the solution**, add 6 M HNO_3 dropwise to each test tube. Watch for a strong evolution of gas (bubbles) as you add the HNO_3. Record your results for the known and the unknown.

$$CO_3^{2-} + 2\,HNO_3 \longrightarrow \underset{bubbles}{CO_2(g)} + H_2O + 2\,NO_3^-$$

EXPERIMENT 24

C. DETERMINATION OF THE FORMULA OF AN UNKNOWN SALT

Your unknown consists of a salt composed of a cation and an anion. The tests you performed on the unknown will give you information about the ions. From your results, determine the presence of one of the cations tested (sodium, potassium, calcium, iron or ammonium), and one of the anions (chloride, phosphate, sulfate, or carbonate). For example, if your unknown salt was $CaCl_2$, you would have seen a positive test for Ca^{2+} and a positive test for Cl^- as you tested the unknown.

C.1   Write the formulas of the cation and anion (one of each) you identify as present in your unknown.

C.2   Use the ionic charges of the cation and anion to write the formula and name of the salt that was your unknown.

EXPERIMENT 24
TESTING FOR ANIONS AND CATIONS
LABORATORY REPORT

A. TESTS FOR CATIONS

Ion tested	Test	Observations
A.1 Na^+	flame test	
A.2 K^+	flame test	
A.3 Ca^{2+}	flame test	
	oxalate	
A.5 Fe^{3+}	KSCN	
A.6 NH_4^+	NaOH, heat	

Unknown Solution   **Sample Number** _____

Unknown solution	Test	Observations	Present or absent?
A.4 Na^+, K^+	flame test		
A.4 Ca^{2+}	flame test		
	oxalate		
A.5 Fe^{3+}	KSCN		
A.6 NH_4^+	NaOH, heat		

EXPERIMENT 24

B. TESTS FOR ANIONS

B.1 Silver nitrate test

Known Ions	Observations	
	AgNO$_3$	adding HNO$_3$ to precipitate
Cl$^-$		
PO$_4^{3-}$		
SO$_4^{2-}$		
CO$_3^{2-}$		

Unknown solution	Observations		Ion(s) present
	AgNO$_3$	adding HNO$_3$	

B.2 Test for phosphate ion, PO$_4^{3-}$ (Optional)

Known solution	Observations with (NH$_4$)$_2$MoO$_4$
PO$_4^{3-}$	

Unknown solution	Observations with (NH$_4$)$_2$MoO$_4$	PO$_4^{3-}$ present or absent

NAME_____ SECTION_____ DATE_____

B.3 Barium chloride test

Known Ions	Observations	
	$BaCl_2$	adding HNO_3 to precipitate
Cl^-		
PO_4^{3-}		
SO_4^{2-}		
CO_3^{2-}		

Unknown solution	Observations		Ion(s) present or absent
	$BaCl_2$	adding HNO_3 to precipitates	

B.4 *Carbonate ion, CO_3^{2-}*

Known solution	Observations with HNO_3
CO_3^{2-}	

Unknown solution	Observations with HNO_3	CO_3^{2-} present or absent

213

EXPERIMENT 24

C. IDENTIFICATION OF AN UNKNOWN SALT AND ITS FORMULA

Unknown Sample Number _____

C.1   Cation present in unknown   _____

What test results led to this conclusion?   Explain.

Anion present in unknown   _____

What test results led to this conclusion?   Explain.

C.2   Formula of the salt   _____

Name of the salt   _____

NAME_____ SECTION_____ DATE_____

QUESTIONS AND PROBLEMS

1. Write a net ionic equation for the following reactions:

   a. silver nitrate, $AgNO_3$, and NaCl

   b. silver nitrate, $AgNO_3$, and Na_2CO_3

   c. barium chloride, $BaCl_2$, and Na_2CO_3

2. State the cation(s) or anion(s) that give(s) the following reaction:

   _____ a. forms a precipitate with $AgNO_3$ that does not dissolve in HNO_3

   _____ b. forms a gas with HCl

   _____ c. gives a bright, yellow-orange flame test

   _____ d. forms a precipitate with $BaCl_2$ that does not dissolve in HNO_3

   _____ e. forms a deep red color with KSCN

3. For each combination, write the formula of the salt and indicate whether the salt is soluble (S) or insoluble (IS):

CATIONS   ANIONS

	Cl^-	SO_4^{2-}	CO_3^{2-}	PO_4^{3-}
Na^+				
K^+				
Ca^{2+}				
Ba^{2+}			$BaCO_3$ (IS)	
Fe^{3+}				
NH_4^+				

EXPERIMENT 25
ACIDS, BASES pH AND BUFFERS

GOALS

1. Prepare a naturally occurring dye to use as a pH indicator.
2. Use a pH meter and an indicator to determine the pH of several substances.
3. Observe the changes in pH as acid or base is added to buffered and unbuffered solutions.
4. Calculate the [H^+] or the [OH^-] of a solution using the K_w of water.
5. Calculate pH from the [H^+] of a solution.

MATERIALS NEEDED

Bring from home:
Samples to test for pH (shampoo, conditioner, mouth wash, antacids, etc.)
Part A
red cabbage leaves
400 mL beaker
distilled water
test tubes
set of buffers pH 1-13

Part B
shell vials
samples to test
pH meter

Part C
buffer with a high pH, low pH
0.1 M NaCl
0.1M HCl, 0.1M NaOH
pH meter

CONCEPTS TO REVIEW

acids
bases
pH
K_w
buffers

EXPERIMENT 25

BACKGROUND DISCUSSION

An *acid* is a substance that dissolves in water and produces hydrogen ions, H^+. In the laboratory we have been using acids such as hydrochloric acid (HCl), nitric acid (HNO_3), and acetic acid ($HC_2H_3O_2$). Acetic acid is the acid in vinegar that gives it a sour taste.

$$HCl + H_2O \longrightarrow \underset{\text{hydronium ion}}{H_3O^+ + Cl^-}$$

or simplified

$$HCl \xrightarrow{H_2O} H^+ + Cl^-$$

A *base* is a substance that dissolves in water and produces hydroxide ions. Some typical bases in the laboratory are sodium hydroxide (NaOH) and potassium hydroxide (KOH).

$$NaOH \longrightarrow Na^+ + OH^-$$

An important weak base found in the laboratory and in some household cleaners is ammonia. In water, it reacts to form ammonium and hydroxide ions:

$$NH_3 + H_2O \rightleftarrows NH_4^+ + OH^-$$

The concentration (moles/liter) of H^+ or OH^- can be determined from the constant for water:

$$K_w = [H^+][OH^-] = [1.0 \times 10^{-7}][1.0 \times 10^{-7}] = 1.0 \times 10^{-14}$$

If the $[H^+]$ or $[OH^-]$ for an acid or a base is known, the other can be calculated. For example, an acid has a $[H^+] = 1.0 \times 10^{-4}$ M. We can find the $[OH^-]$ of the solution, by setting up the K_w expression for $[OH^-]$:

$$[H^+][OH^-] = 1.0 \times 10^{-14}$$

$$[OH^-] = \frac{1.0 \times 10^{-14}}{[H^+]} = \frac{1.0 \times 10^{-14}}{1.0 \times 10^{-4}} = 1.0 \times 10^{-10} \text{ M}$$

The pH of a solution is a measure of the acidity, basicity, or neutrality of that solution. On the pH scale, pH values below 7 are acidic, equal to 7 is neutral, and values above 7 are basic. Typically, the pH scale has values between 0 and 14.

pH scale

0 1 2 3 4 5 6 7 8 9 10 11 12 13 14
very acidic   *neutral*   *very basic*

The pH of a solution is a measure of its $[H^+]$. It is defined as the negative log of the hydrogen ion concentration.

$$pH = -\log[H^+]$$

Therefore, a solution with a $[H^+] = 1.0 \times 10^{-2}$ has a pH of 2, and is acidic. A solution with a $[H^+] = 1.0 \times 10^{-11}$ has a pH of 11, and is basic.

Many natural substances contain pigments that produce distinctive colors at different pH values. In this experiment we will prepare a solution of the pigment found in red cabbage leaves. By extracting(removing) this pigment from the leaves of the red cabbage to form an aqueous solution, we can prepare a natural indicator which produces many different colors in acidic and basic solutions.

Buffer solutions of known pH values will be set up as a pH reference set from pH 1 to 13. Although the pH scale includes pH 0 and 14, these pH values are not typically found among naturally occurring substances. Adding the red cabbage solution to each will produce a series of distinctive colors. When the red cabbage solution is added to a sample, the color will be matched to the colors of the pH reference set to determine the pH of the sample. A pH meter may also be available for use in the laboratory. Your instructor will give directions for its use. This is the laboratory method for determining the pH of a solution. Then pH values for various samples will obtained by both the indicator methods, and by using the pH meter.

The pH of the blood is maintained between 7.35 and 7.45 by buffers in the body. Buffers keep pH constant because they are able to react with small amount of acids or bases that are added to the solution. A typical buffer may be composed of a weak acid and its salt. The weak acid reacts with base, and the anion of the salt picks up H^+. In this way, the effects of compounds in the body that would change the pH of the blood are offset by the buffer system, and a specific pH is maintained. For example, the bicarbonate buffer system in the blood contains the weak acid, carbonic acid, H_2CO_3, and bicarbonate anion, HCO_3^-. When base (OH^-) is added, it reacts with the weak acid to produce the bicarbonate ion and water:

$$H_2CO_3 + OH^- \longrightarrow HCO_3^- + H_2O$$

When acid enters the blood, the H^+ reacts with the HCO_3^- anion and forms carbonic acid:

$$HCO_3^- + H^+ \longrightarrow H_2CO_3$$

In the buffer section of this experiment, you will observe how the pH of a buffered and unbuffered solution is affected by the addition of acid and base.

LABORATORY ACTIVITIES

A. COLORS OF A pH REFERENCE SET USING RED CABBAGE INDICATOR

Tear up several leaves of a red cabbage and place them in a 400 mL beaker. Add about 100-150 mL of distilled water to cover the leaves. Set the beaker on a wire gauze on an iron ring attached to the ring stand. Heat the water, but do not let it boil vigorously. In 5-10 min, the solution should become a purple color. If not, add another 1-2 leaves of red cabbage, and heat again. Let the red cabbage solution cool.

While the pigment in the red cabbage is being extracted, prepare your pH reference set of pH buffers. The red cabbage solution will be needed for the other parts of this experiment.

EXPERIMENT 25

Preparation of a pH reference set

Your instructor may have the class prepare one pH scale reference set. To set up the pH reference set, you need 13 test tubes. This may be accomplished by combining your test tube set with a partner's set. Place 3-4 mL of a buffer from pH 1-13 in each of the test tubes. You should now have test tubes arranged in order of pH 1 through 13.

To each test tube and buffer, add 2-3 mL (40-60 drops) of the red cabbage solution. If you wish a deeper color, add another 1 mL of the indicator. Record the colors of the pH solutions. **Keep** this pH reference set for the next part of the experiment.

B. DETERMINATION OF pH IN SAMPLE SOLUTIONS

Place 3-4 mL of one of the samples in a shell vial (or a small beaker). Add 2-3 mL (40-60 drops) of red cabbage solution. Take the sample to the pH reference set, and compare the colors. The pH of the buffer in the reference set that has the closest color match is the pH you record for the sample. Continue to test other samples in the same way. Record your results.

pH Meter If a pH meter is available, it may be used to determine the pH values of the samples. In addition, samples with strong colors that cannot be used with cabbage indicator, such as coffee and cola, may be checked with the pH meter. After you have determined the pH with the red cabbage solution, take the sample to a pH meter. Calibrate the meter if necessary. Place the electrodes in the sample, and determine the pH. Your instructor will give your directions for your particular pH meters. Record.

C. EFFECT OF BUFFERS UPON pH CHANGE

C.1 Place 5 mL of each of the following solutions into four shell vials (or test tubes):

1. H_2O
2. 0.1 M NaCl
3. A buffer with a high pH
4. A buffer with a low pH

Determine the initial pH of each solution. (Use a pH meter, or add 2 mL of cabbage indicator.) Record.

C.2 Add 5 drops of 0.1 M HCl (acid) to each of the solutions. Stir and determine the pH again of each solution.

C.3 Determine the change in pH units, if any. Record.

C.4 Set up four shell vials (or test tubes) as you did in step C.1 using the same four solutions. Determine the initial pH. (Use a pH meter, or add 2 mL of cabbage indicator to each.)

C.5 Add 5 drops of 0.1 M NaOH (base) to each. Stir and determine the pH again.

C.6 Determine the change in pH units. Record.

NAME_____ SECTION_____ DATE_____

EXPERIMENT 25
ACIDS, BASES, PH, AND BUFFERS
LABORATORY REPORT

A. COLORS OF A pH REFERENCE SET USING RED CABBAGE INDICATOR

pH	Acidic	pH	Basic
1	_____	8	_____
2	_____	9	_____
3	_____	10	_____
4	_____	11	_____
5	_____	12	_____
6	_____	13	_____

neutral 7 _____

Complete the following table:

$[H^+]$	$[OH^-]$	pH	acidic, basic, neutral?
1×10^{-6}	_____	_____	_____
_____	_____	10	_____
_____	1×10^{-3}	_____	_____
_____	_____	_____	Neutral

EXPERIMENT 25

B. DETERMINATION OF pH

Solution	Color with indicator	pH	pH by pH meter	Acidic, basic, or neutral?
household cleaners				
vinegar	_____	_____	_____	_____
ammonia	_____	_____	_____	_____
_____	_____	_____	_____	_____
drinks, juices				
lemon juice	_____	_____	_____	_____
apple juice	_____	_____	_____	_____
_____	_____	_____	_____	_____
detergents, shampoos				
detergent	_____	_____	_____	_____
dish soap	_____	_____	_____	_____
shampoos	_____	_____	_____	_____
_____	_____	_____	_____	_____
health aids				
mouth wash	_____	_____	_____	_____
antacid	_____	_____	_____	_____
aspirin	_____	_____	_____	_____
other items tested				
_____	_____	_____	_____	_____
_____	_____	_____	_____	_____

NAME_____ SECTION_____ DATE_____

C. EFFECT OF BUFFERS UPON pH CHANGE

Effect of adding 0.1 M HCl

	Initial pH (C.1)	Final pH (C.2)	pH change (C.3)
H_2O			
0.1 M NaCl			
buffer 1			
buffer 2			

Effect of adding 0.1 M NaOH

	Initial pH (C.4)	Final pH (C.5)	pH change (C.6)
H_2O			
0.1 M NaCl			
buffer 1			
buffer 2			

Which solution(s) showed the greatest change of pH? Why?

Which solutions(s) showed little or no change in pH? Why?

What is the function of a buffer?

EXPERIMENT 25

QUESTIONS AND PROBLEMS

1. The label on the shampoo claims that it is pH balanced. What do you think "pH balanced" means?

2. A solution has a $[OH^-] = 1.0 \times 10^{-5}$ M. What is the $[H^+]$ and the pH of the solution?

3. A sample of 0.0020 mole HCl is dissolved in water to make a 2000 mL solution. Calculate the molarity of the HCl solution; the $[H^+]$, and the pH. (Hint: Since HCl is a strong acid, HCl $\longrightarrow$ H$^+$ + Cl$^-$, the $[H^+]$ can be found from the molarity of the HCl solution.)

4. Normally, the pH of the human body is fixed in a very narrow range between 7.35 and 7.45 A patient with a acidotic blood pH of 7.3 may be treated with alkali such as sodium bicarbonate. Why would this treatment raise the pH of the blood?

EXPERIMENT 26
ACID BASE TITRATIONS

GOALS

1. Prepare a sample for titration with a base.
2. Set up a buret and use proper titration technique in reaching an endpoint.
3. Calculate the concentration and percentage of acetic acid in vinegar.
4. Determine the acid-absorbing capacity of a commercial antacid.

MATERIALS NEEDED

vinegar (white)
250 mL Erlenmeyer flask
10 ml graduated cylinder
  or a 5 mL pipet and bulb
phenolphthalein indicator
50 mL buret
buret or butterfly clamp
small funnel

antacid samples
0.10 M HCl (standardized)
0.10 M NaOH (standardized)
bromphenol blue indicator
mortar and pestle

CONCEPTS TO REVIEW

acid
base
neutralization
titration
molarity
percent weight/volume

BACKGROUND DISCUSSION

Vinegar is an aqueous solution of acetic acid ($HC_2H_3O_2$, CH_3COOH, or simply HAc). The concentration (molarity and percent) of an acidic solution can be determined by using a *neutralization* reaction of an acid and a base. Since acetic acid has one acidic hydrogen per mole, one mole of acetic acid will be neutralized by one mole of NaOH. The products of neutralization are a salt and water.

$$CH_3COOH \ + \ NaOH \longrightarrow CH_3COO^- Na^+ \ + \ H_2O$$
acetic acid   base   salt

EXPERIMENT 26

In a *titration*, we carry out a neutralization reaction in the presence of an indicator, such as phenolphthalein. When all the acid has been neutralized by an added volume of NaOH, the indicator phenolphthalein will turn from a clear color to a faint, but permanent pink color. This indicates that you have reached an *endpoint*. When you reach this point, stop adding base. A final reading of the buret is taken to determine the volume of NaOH required to reach the endpoint.

Calculation of molarity of acetic acid

By knowing the volume of the NaOH added, and its molarity from the label on the NaOH reagent bottle, we can calculate the moles of NaOH that were required to neutralize the acid.

$$\text{moles NaOH} = \text{L NaOH} \times \frac{\text{moles NaOH}}{\text{L NaOH}}$$

In this neutralization reaction, the moles of NaOH added are equal to the moles of acetic acid (HAc) present in the sample.

$$\text{NaOH} + \text{HAc} \longrightarrow \text{NaAc} + \text{H}_2\text{O}$$

moles HAc = moles NaOH

Using the moles of HAc, we can calculate the molarity of HAc in the 5.0 mL sample of vinegar. Since the volume must be expressed in liters, we write the 5.0 mL volume as 0.0050 L.

$$\text{molarity (M) HAc} = \frac{\text{moles HAc}}{0.0050 \text{ L HAc}}$$

Typically, a titration is repeated with three samples. The molarity for each is calculated, and then an average molarity is determined.

Calculation of the percent (weight/volume) of acetic acid

To calculate the percent (weight/volume) of the acetic acid, we convert the moles HAc to grams using the molar mass of acetic acid, 60.0 g/mole.

$$\text{g HAc} = \text{moles HAc} \times \frac{60.0 \text{ g HAc}}{1 \text{ mole HAc}}$$

$$\text{percent (weight/volume)} = \frac{\text{g HAc}}{5.0 \text{ mL}} \times 100$$

Absorbing capacity of commercial antacids

Stomach acid is primarily hydrochloric acid (HCl) which has a concentration of about 0.1 M. Sometimes, when a person is under stress, excess HCl may be produced, causing discomfort. An agent called an **antacid** is used to absorb some of the acid. One commercial claims that its antacid product consumes 47 times its own weight in excess stomach acid; 1 g of the antacid neutralizes 47 g of HCl in the stomach.

In this experiment, we will test such a claim by weighing an antacid tablet and reacting that tablet with a solution of 0.1 M HCl. The amount of HCl will exceed the amount neutralized by the antacid which leaves some HCl unreacted. A *back titration* with NaOH will give the amount of acid that was *not* absorbed by the antacid. This amount is subtracted from the 100 mL of HCl we added initially to give the amount of HCl that was absorbed. Several commercial antacid products may be tested.

LABORATORY ACTIVITIES

Wear your safety glasses!

A. CONCENTRATION OF ACETIC ACID IN VINEGAR

A.1 Using a 10 ml graduated cylinder or a 5.0 mL pipette, transfer 5.0 mL of vinegar to a 250 mL Erlenmeyer flask. See Figure 26.1 for use of a pipet.

 If you are using a pipet, use a suction bulb to draw vinegar into a pipet. Do not mouth pipet. It is dangerous!

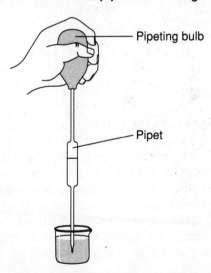

Figure 26.1 *Using a bulb to draw a liquid into a pipet. While liquid is above the volume line, remove the bulb and quickly place index figure over the end of the stem. Slowly lower the liquid to the line, and then transfer to an Erlenmeyer flask. A small amount remaining in the tip has been included in the calibration of the pipet.*

EXPERIMENT 26

Record the brand of vinegar and the % acetic acid stated on the label. Record the volume of the vinegar sample. Add about 25 mL of water to increase the volume of the solution for titration. This will not affect your results. Add 2-3 drops of the phenolphthalein indicator to the solution in the flask.

A.2 Using a 250 mL beaker, obtain about 200 mL of NaOH solution. Record the *molarity* of the NaOH solution as stated on the label of the reagent bottle.

Obtain a buret from the stockroom. Clean the buret by pouring through several portions of water from a beaker. If a detergent and buret brush are available, clean the buret with detergent. Rinse several times, and finally rinse with distilled water. Then pour two 5 mL portions of the NaOH solution through the buret. Discard the NaOH washings.

Using a buret or butterfly clamp, set up the buret as demonstrated by your instructor. See Figure 26.2.

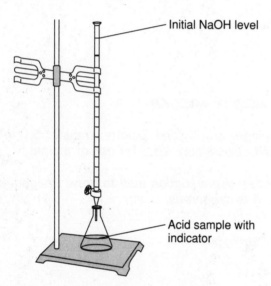

Figure 26.2 *Buret setup for acid-base titration*

A.3 Observe that the buret is marked 0.0 mL at the top of the buret and 50.0 mL at the bottom. Volume can be read to the 0.1 mL. (You can estimate to the nearest 0.01 mL.) To fill the buret with NaOH, place a small funnel in the top of the buret. Carefully pour NaOH from the beaker into the buret. Lift the funnel as you reach the top, so the NaOH does not spill over. Add NaOH until it is *above* the 0.0 mL line. Adjust the meniscus to the top line (0.0 mL) by draining some NaOH into an extra beaker. The buret tip should now be full of NaOH solution, and free of bubbles. Discard the NaOH removed during this adjustment. Record the initial buret reading of NaOH (usually as 0.0 mL at the top of the buret).

A.4 Place the flask containing the vinegar solution under the buret on a piece of white paper. Begin to add NaOH to the solution. (Be sure you have added indicator.) This is done by opening and closing the stopcock with your left hand (if you are right handed), and swirling the flask with your right hand to mix the acid and the base.

At first, the pink streaks produced by the reaction will disappear quickly. However, as you approach the end point, the pink streaks will be more persistent and disappear slowly. **Slow** down the addition of the NaOH to drops at this time. *Soon, one drop of NaOH will cause the entire solution to turn a faint, permanent pick color.* **STOP ADDING NaOH**. You have reached the end point of the titration. See Figure 26.3.

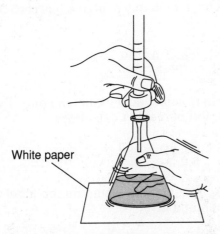

Figure 26.3 *The solution in the flask is swirled as NaOH is added to the sample during titration.*

At the endpoint, the acid in the sample has been neutralized. Record the final buret reading of the NaOH. Then fill the buret again, and repeat the titration two more times with samples of the same type of vinegar.

Calculations:

A.5 Calculate the volume of NaOH used to neutralize the vinegar samples (A.4 - A.3). When two or three samples of vinegar are titrated, calculate the average volume of NaOH used. Total the volumes of NaOH used, and divide by the number of samples you titrated.

$$\frac{\text{Volume (1)} + \text{Volumes (2)} + \text{Volume (3)}}{3}$$

Calculation of molarity

A.6 Convert the volume (average) of NaOH used to a volume in liters (L).

$$L = mL \times \frac{1 \, L}{1000 \, mL}$$

A.7 Calculate the moles of NaOH using the volume (L) and molarity of the NaOH.

$$\text{moles NaOH} = L \, \text{NaOH} \times \frac{\text{moles NaOH}}{L \, \text{NaOH}}$$

EXPERIMENT 26

A.8 Record the moles of acid present in the vinegar which are equal to the number of moles of NaOH in A.7.

A.9 Calculate the molarity(M) of the acetic acid in the vinegar sample (A.8÷A.6).

$$\text{molarity (M) HAc} = \frac{\text{moles HAc}}{0.0050 \text{ L HAc}}$$

Calculation of percent (weight/volume)

A.10 Using the number of moles in A.8, calculate the number of grams of acetic acid (HAc) in the vinegar sample. The molar mass of HAc (CH_3COOH) is 60.0 g/mole.

$$\text{g HAc} = \text{moles HAc (A.8)} \times \frac{60.06 \text{ g HAc}}{1 \text{ mole HAc}}$$

A.11 Calculate the percent (weight/volume) acetic acid in the vinegar. For an original volume of 5.0 mL vinegar, the percent is calculated.

$$\text{percent (weight/volume)} = \frac{\text{g HAc} \times 100}{5.0 \text{ mL}}$$

A.12 From you observation of the %HAc (acetic acid) listed on the label of the vinegar bottle, calculate your percentage error.

$$\text{percent error} = \frac{|\text{experiment \% - label \%}|}{\text{label \%}} \times 100$$

B. TITRATION OF AN ANTACID

B.1 Record the name of the antacid and its active ingredient listed on the label.

B.2 Weigh a 250 mL Erlenmeyer flask to the nearest 0.01 g. Record its mass. Crush an antacid tablet, and transfer the crushed tablet to the Erlenmeyer flask. Weigh the flask and the crushed antacid. Record. This gives you the mass of antacid you have in the flask.

B.3 Obtain 100 mL of standardized HCl. Record the given molarity on the reagent bottle (it should be close to 0.1 M). Add the 100 mL of HCl to the flask. Using a Bunsen burner or a hot plate, bring the solution to a boil for 5 minutes. The solution may be cloudy because starches and binders are not very soluble. Add 4-5 drops of bromphenol blue indicator. The solution should be yellow. If it turns blue-green, add another 25 mL of the 0.1 M HCl. Record the total volume of HCl added.

B.4 The antacid will react with and neutralize a certain amount of the HCl. To determine this amount, we need to find out how much HCl remains that did not react with the antacid. Fill a buret with NaOH to a level slightly above the 0.0 mL reading. Let the excess run out into a separate beaker to bring the volume of NaOH down to the 0.0 mL mark. Record the initial buret reading of the NaOH.

ACID-BASE TITRATIONS

B.5 Titrate the unreacted HCl remaining in the antacid solution with NaOH. The endpoint is reached when the yellow color of the solution turns to a blue-green color. Record the final buret reading of the NaOH.

Calculations

B.6 Determine the mass of the antacid.

B.7 Calculate the volume of NaOH used in the titration (B.5-B.4).

B.8 Calculate the volume of unreacted HCl that was titrated with NaOH. This will be equal to the volume of NaOH (B.7), if the molarities of HCl and NaOH are the same.

B.9 Calculate the volume of HCl that reacted with the antacid.

HCl reacted = mL HCl added (B.3) - volume(mL) of unreacted HCl(B.8)

B.10 If we assume that 1 mL of the 0.1 M HCl has the same density as water (1.0 g/mL), we can convert the mL of HCl solution to grams of HCl that reacted with the antacid.

$$\text{mL HCl reacted} \times \frac{1 \text{ g}}{1 \text{ mL}} = \text{g HCl reacted}$$

B.11 The mass of HCl absorbed by 1 g of the antacid can now be determined.

$$\frac{\text{g HCl reacted}}{\text{g antacid sample}} = \text{g acid absorbed by 1 g of antacid}$$

B.12 Record values obtained in the class for other antacids.

EXPERIMENT 26
ACID-BASE TITRATIONS
LABORATORY REPORT

A. CONCENTRATION OF ACETIC ACID IN VINEGAR

A.1 Brand of vinegar   _____

% acetic acid (label)   _____%

	Sample 1	Sample 2	Sample 3
Volume (mL) of vinegar	_____mL	_____mL	_____mL

A.2 M of NaOH (from label)   _____M

A.4 Final buret reading   _____mL   _____mL   _____mL

A.3 Initial buret reading   _____mL   _____mL   _____mL

A.5 Volume (mL) of NaOH used   _____mL   _____mL   _____mL

A.6 Average volume (mL) of NaOH   _____mL

Average volume (L) of NaOH   _____L

A.7 Moles NaOH   _____mole NaOH
Show calculations:

A.8 Moles HAc   _____mole HAc

Molarity calculations:

A.9 M of HAc   _____ M (moles/L)
Show calculations:

EXPERIMENT 26

Percent (weight/volume) concentration

A.10 grams HAc _____ g
       Show calculations:

A.11 % HAc in vinegar _____ %
       Show calculations:

A.12 % Error _____ %
       Show calculations:

NAME_____ SECTION_____ DATE_____

B. TITRATION OF AN ANTACID

B.1 Brand of antacid _____ g

     Active ingredient(s) _____

B.2 Mass of flask and antacid _____ g

     Mass of flask _____ g

B.3 Volume of
0.10 M HCl added _____ mL

B.5 Final buret reading _____ mL

B.4 Initial buret reading _____ mL

     M NaOH _____ M

Calculations

B.6 Mass of antacid _____ g

B.7 Volume of
NaOH used _____ mL

B.8 Volume of
unreacted HCl _____ mL

B.9 Volume of
reacting HCl _____ mL

B.10 g HCl absorbed _____ g

B.11 g HCl absorbed by
1 g antacid _____ g/1 g antacid
Show calculations:

B.12   *Summary of Absorbing Power of Various Antacids*

    Antacid   Mass HCl absorbed per 1 gram antacid

    _____   _____

    _____   _____

    _____   _____

EXPERIMENT 26

QUESTIONS AND PROBLEMS

1. What are the reactants in a neutralization reaction?

   What are the products?

2. Write a balanced reaction for the following neutralization reactions:

   sodium hydroxide and hydrochloric acid

   potassium hydroxide (KOH) and sulfuric acid (H_2SO_4)

   lithium hydroxide and phosphoric acid (H_3PO_4)

3. How many mL of 0.10 M NaOH will be required to completely react with 15.0 mL of 0.200 M H_2SO_4?

   H_2SO_4 + 2 NaOH $\longrightarrow$ Na_2SO_4 + 2 H_2O

4. A solution contains 15.0 g of KCl in 75.0 mL of solution.

   a. What is the percent (w/v) concentration of the KCl solution?

   b. What is the molarity of the KCl solution?

5. How many moles of NaOH are needed to prepare 2.00 L of a 0.250 M NaOH solution?

EXPERIMENT 27
COMPARING PROPERTIES OF ORGANIC AND INORGANIC COMPOUNDS

GOALS

1. To compare bonding, solubilities, densities, flammability, melting and boiling points of organic and inorganic compounds.
2. To list physical characteristics most typical of hydrocarbons and inorganic compounds.

MATERIALS NEEDED

evaporating dish (2)
test tubes
chemicals: pentane, cyclohexane, NaCl(s)
chemistry handbook(s)

CONCEPTS TO REVIEW

carbon and covalent compounds
properties of organic compounds
solubility

BACKGROUND DISCUSSION

A group of compounds known as *organic compounds* share certain properties that can be used to distinguish them from the inorganic compounds we have studied. Many organic compounds called hydrocarbons contain covalent bonds between the elements carbon and hydrogen. Other families of organic compounds also contain oxygen and/or nitrogen.

In this experiment, we will study some properties of organic compounds and compare them to the properties of inorganic substances. These properties include color of the substances, physical state, density, solubility, melting and boiling points, and flammability. Most organic compounds are nonpolar, a feature that makes them soluble in nonpolar solvents, and not in water. The covalent bonds of organic compounds give them melting and boiling points which are generally lower than those of polar and ionic compounds. Organic compounds are very volatile and flammable. Many such as propane are used as fuels and undergo combustion with oxygen in air easily.

$$C_3H_8 + 5\,O_2 \xrightarrow{\text{heat}} 3\,CO_2 + 4\,H_2O$$
propane

EXPERIMENT 27

LABORATORY ACTIVITIES

 ORGANIC COMPOUNDS ARE EXTREMELY FLAMMABLE! USE OF THE BUNSEN BURNER WILL BE PROHIBITED. DISPOSE OF ORGANIC COMPOUNDS IN SPECIALLY MARKED CONTAINERS AS INSTRUCTED BY YOUR INSTRUCTOR.

 WEAR YOUR SAFETY GLASSES!

A. COLOR AND PHYSICAL STATE

Describe the color and physical state (solid, liquid, or gas) of pentane and sodium chloride. Write a formula for each.

B. COMBUSTION

This will be a demonstration by your instructor.

Place 10 drops of pentane, and a small amount of NaCl in separate evaporating dishes. Test their flammability by using a lighted match or splint to try to ignite the compound. Record your observations.

C. SOLUBILITY AND DENSITY

Test the solubility of pentane and NaCl, in water, a polar solvent, and in cyclohexane, a nonpolar solvent. Place 2-3 mL of water in two test tubes. Add 10 drops of pentane to the first test tube. Add a few crystals of NaCl(s) to the second test tube. Stir the contents of each test tube. Record your observations of each substance as soluble (S) or insoluble (I) in water. If one of the substances is not soluble, indicate whether it is less dense than water (it will float), or more dense (it will sink).

Repeat the above procedure using 2-3 mL of cyclohexane as the solvent. Record your observations. *Dispose of the organic substances in the proper waste container.*

Using a chemistry hand book, look up and record the densities of pentane and sodium chloride.

D. MELTING AND BOILING POINTS

Using a chemistry handbook, look up the melting and boiling points of pentane and NaCl. Record.

E. SUMMARY OF PROPERTIES

Complete the table summarizing the properties of organic and inorganic compounds using pentane as a typical organic compound and NaCl as a typical inorganic compound.

NAME_____ SECTION_____ DATE_____

EXPERIMENT 27
COMPARING PROPERTIES OF ORGANIC AND INORGANIC COMPOUNDS
LABORATORY REPORT

A. COLOR AND PHYSICAL STATE

	Pentane	Sodium chloride
Color	_____	_____
State	_____	_____
Formula	_____	_____

B. COMBUSTION

Which substance was flammable? _____

Write an equation for the combustion of that compound.

C. SOLUBILITY AND DENSITY

	Pentane	Sodium chloride
Solubility		
water	_____	_____
cyclohexane	_____	_____
Sink or float	_____	_____
Density (handbook)	_____	_____

EXPERIMENT 27

D. MELTING AND BOILING POINTS

	Pentane	Sodium chloride
melting point	_____	_____
boiling point	_____	_____

E. SUMMARY OF PROPERTIES

	Pentane (organic)	Sodium chloride (inorganic)
Type of compound	_____	_____
Elements present	_____	_____
Bonding	_____	_____
Flammable	_____	_____
Solubility	_____	_____
Density	_____	_____
Melting point	_____	_____
Boiling point	_____	_____

EXPERIMENT 28
SATURATED HYDROCARBONS

GOALS

1. Construct molecular models of alkanes.
2. Write the names of alkanes from their structural formulas.
3. Write the complete structural and condensed structural formulas of the isomers having the same molecular formula.

MATERIALS NEEDED

molecular model kit
chemistry handbook

CONCEPTS TO REVIEW

alkanes
cycloalkanes
haloalkanes
structural formulas
condensed formulas
isomers
naming alkanes

BACKGROUND DISCUSSION

The saturated hydrocarbons represent a group of organic compounds composed of the elements carbon and hydrogen. In this experiment, we will study the alkanes and cycloalkanes which are also called *saturated* hydrocarbons because they contain only single bonds. We will also look at their halogen derivatives, the haloalkanes. In each type of alkane, each carbon atom has four valence electrons and must always have four single bonds to other carbon, hydrogen or halogen atoms.

To learn more about the bonding of alkanes, it is helpful to build models using a molecular model kit. In the kit, there are wooden (or plastic) balls which represent the elements. Typically, carbon is black, oxygen is red, hydrogen is yellow, and nitrogen is blue. Each of the wooden atoms have the correct number of holes drilled for bonds that attach to other atoms: Carbon has four bonds, hydrogen has one bond, oxygen had two bonds, and nitrogen has three bonds. Each ball-and-stick model of a hydrocarbon compound must follow these bonding requirements.

EXPERIMENT 28

The first ball-and-stick model to construct is methane, CH_4. When you build the methane model, you will notice that it has a three-dimensional shape called a tetrahedron. We can draw this three-dimensional structure, but it becomes complex as the carbon chain gets longer. To represent this structure on paper, we mentally flatten out its shape, and write the carbon atom attached by four bonds to four hydrogen atoms. This type of formula is called a *complete structural formula*.

Three-dimensional structure Complete structural formula   Condensed structural formula
(tetrahedron)

As the number of carbon atoms in a hydrocarbon increases, the writing of the complete structural formula becomes time consuming. For convenience, chemists use a shortened version called a *condensed structural formula*. To convert a structural formula to a condensed formula, we group the hydrogen atoms attached to each carbon atom. The number of hydrogen atoms are shown as a subscript following the carbon atom. Even though hydrogen atoms appear between the carbon atoms in the condensed version, we understand from the models of the compound that the backbone of the structure is the chain of the carbon atoms.

Structural formulas   Condensed formulas

Isomers

When a molecular formula represents a group of atoms that can be arranged in two or more different ways, different compounds called *isomers* are obtained. Isomers are present when a molecular formula can represent two or more different structural (or condensed) formulas. One structure cannot be converted to the other without breaking and forming new bonds. The isomers have different physical and chemical properties. One of the reasons for the vast array of organic compounds is the phenomenon of isomerism.

Cycloalkanes

In a cycloalkane, an alkane has a cyclic or ring structure. There are no end carbon atoms. The structural formula of a cycloalkane indicates all of the carbon and hydrogen atoms. The condensed formula groups the hydrogen atoms with each of the carbon atoms. Another type of notation called the *geometric* structure is often used to depict a cycloalkane by showing only the bonds that outline the geometric shape of the compound. For example, the geometric shape of cyclopropane is a triangle, while the geometric shape of cyclobutane is a square. Examples of the various structural formulas for cyclobutane are shown below.

Structural formula   Condensed formula   Geometric formula

Haloalkanes

The haloalkanes are derivatives of alkanes (or cycloalkanes) in which a halogen atom has replaced one or more hydrogen atoms in the alkane.

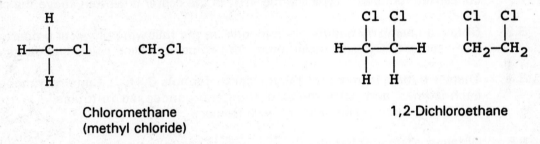

Chloromethane
(methyl chloride)

1,2-Dichloroethane

LABORATORY ACTIVITIES

A. STRUCTURE OF ALKANES

A.1   The hydrocarbons called the **alkanes** contain only single bonds. Using an organic model kit, identify the color of the carbon atoms, and the hydrogen atoms. The number of holes drilled in the wooden balls represent the number of covalent bonds that are formed by that atom. The location of the holes represents the actual three-dimensional angles of the bonds. Write the electron dot structures for an atom of carbon and an atom of hydrogen. State the number of bonds required by carbon and hydrogen.

A.2   Write the electron dot structure of methane, an alkane with one carbon atom, CH_4.

A.3   Prepare a ball-and-stick model of methane by using the small wooden dowels to attach hydrogen atoms to the carbon atom. Draw, as best you can, the tetrahedral shape of methane. Write the complete structural formula of methane. Draw its condensed structural formula by writing the number of hydrogen atoms attached as a subscript.

EXPERIMENT 28

A.4 Ethane is an alkane with two carbon atoms, and propane has three carbon atoms. Write the electron dot structure for ethane and propane.

A.5 Prepare models of ethane and propane. Observe the tetrahedral shape about each carbon atom in the structures. Write the complete structural and condensed structural formulas for each.

B. ISOMERS

B.1 The molecular formula of butane is C_4H_{10}. Construct a model of butane, an alkane with four carbon atoms in a continuous chain. Write its structural and condensed formulas. This model represents a straight-chain isomer.

B.2 Using the model you prepared in B.1, remove a -CH_3 group from the end of the chain. Remove a hydrogen atom from the center carbon atom, and attach the -CH_3 group. Complete the end of the chain with a hydrogen atom. This structure called 2-methylpropane is a **branched-chain** isomer. It is an isomer because its molecular formula is the same as butane, C_4H_{10}. Write its structural and condensed formulas. Typically the -CH_3 in the center is written above the chain.

B.3 Obtain a chemistry handbook, and look up the following physical properties of each isomer of C_4H_{10}: molar mass, melting point, boiling point, and density.

B.4 There are three isomers for the molecular formula C_5H_{12}. Construct models of each isomer, and write the structural and condensed formulas. Write the condensed formula and name of each isomer.

B.5 Obtain a chemistry handbook, and look up the following physical properties of each isomer of C_5H_{12}: molar mass, melting point, boiling point, and density.

C. CYCLOALKANES

C.1 Use the springs in the model kits to connect the carbon atoms in the preparation of models of cycloalkanes with three carbon atoms, four carbon atoms, and five carbon atoms. Write the full and condensed structural formulas. Write the geometric formula for each.

C.2 Name each of the cyclic compounds.

D. HALOALKANES

D.1 We have seen that the haloalkanes are derivatives of alkanes in which a hydrogen atom has been replaced with a halogen atom. Construct models of haloalkanes using the wooden balls that represent chlorine (green), bromine (orange), iodine (violet). Draw the structural and condensed formulas.

D.2 Prepare models of four isomers of dichloropropane. Draw their structural and condensed formulas.

D.3 Give the complete name of each of the compounds.

NAME_____ SECTION_____ DATE_____

EXPERIMENT 28
ALKANES   SATURATED HYDROCARBONS
LABORATORY REPORT

A.   STRUCTURE OF ALKANES

A.1   Color   Electron Dot Structure   Number of bonds

carbon   _black_   _____   _____

hydrogen   _white_   _____   _____

A.2   Electron structure of methane

A.3   Tetrahedron   Structural formula   Condensed formula

A.4   Electron structures

Ethane	Propane

EXPERIMENT 28

A.5

Structural formula	Condensed formula
Ethane	
Propane	

B. ISOMERS

B.1-B.2

C_4H_{10}

Structural formula	Condensed formula
Butane	
2-Methylpropane	

NAME_____ SECTION_____ DATE_____

B.3 Physical Properties

	molar mass	melting point	boiling point	density (g/mL)
Butane	_____	_____	_____	_____
2-Methylpropane	_____	_____	_____	_____

B.4   Structural formula   Condensed formula

Name:	

Name:	

Name:	

B.5 Physical Properties of Isomers of C_5H_{12}

	molar mass	melting point	boiling point	density (g/mL)
_____	_____	_____	_____	_____
_____	_____	_____	_____	_____
_____	_____	_____	_____	_____

EXPERIMENT 28

C. CYCLOALKANES

Structural formula	Condensed formula	Geometric formula
Name:		
Name:		
Name:		

NAME_____ SECTION_____ DATE_____

D. HALOALKANES

D.1   Structural formula   Condensed formula

Chloromethane	

1,2-Dibromoethane	

2-Iodopropane	

EXPERIMENT 28

D.2-D.3 Isomers of dichloropropane

Structural formula	Condensed formula
Name:	
Name:	
Name:	
Name:	

NAME_____ SECTION_____ DATE_____

QUESTIONS AND PROBLEMS

1. Write the correct name of the following alkanes, cycloalkanes, and haloalkanes.

   a. $CH_3CH_2CH_3$ _____

   b. $CH_3\overset{\overset{\displaystyle CH_3}{|}}{C}H\underset{\underset{\displaystyle Cl}{|}}{C}HCH_2CH_3$ _____

   c. $CH_3-\overset{\overset{\displaystyle CH_3}{|}}{\underset{\underset{\displaystyle CH_3}{|}}{C}}-CH_3$ _____

   d. (cyclopentane structure) _____

   e. (bromocyclohexane structure) _____

2. Draw structures for the following alkanes, cycloalkanes and haloalkanes:

   ethane _____

   2,2-dimethylpentane _____

   methylcyclobutane _____

   1,2-dibromocyclohexane _____

251

EXPERIMENT 29
SOME PROPERTIES OF HYDROCARBONS

GOALS

1. Determine the solubility of hydrocarbons in water.
2. Observe the reactions of alkanes and alkenes with bromine and/or potassium permanganate.
3. Write the equations for the reactions of alkanes and/or alkenes that undergo addition, substitution and/or combustion reactions.

MATERIALS NEEDED

organic model kits
test tubes and test tube rack
evaporating dish
2% bromine solution(2% Br_2/1,1,1,-tricholorethane)
0.1 M $KMnO_4$
cyclohexane
cyclohexene
toluene
vegetable oil

CONCEPTS TO REVIEW

saturated, unsaturated and aromatic hydrocarbons
solubility of hydrocarbons
combustion
addition reactions
substitution reactions

BACKGROUND DISCUSSION

Alkanes, *saturated* hydrocarbons, contain single bonds, while the *unsaturated* hydrocarbons of alkenes and alkynes contain double or triple bonds between carbon atoms. The aromatic compounds are hydrocarbons with a benzene ring.

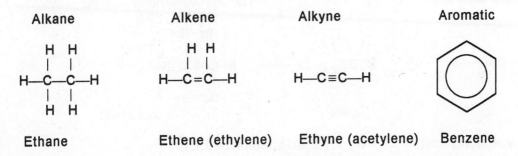

Alkane — Ethane
Alkene — Ethene (ethylene)
Alkyne — Ethyne (acetylene)
Aromatic — Benzene

EXPERIMENT 29

The location of the double or triple bond, an unsaturated site, is a place of reactivity in molecules of alkenes or alkynes.

$H_2C=CH_2$

Ethene (ethylene)

$H_2C=CH-CH_3$

Propene (propylene)

$HC \equiv CH$

Ethyne (acetylene)

$HC \equiv C-CH_3$

Propyne (methyl acetylene)

In this experiment, you will observe some of the physical and chemical properties of saturated and unsaturated hydrocarbons. The solubility of hydrocarbons in water will identify the compounds as polar or nonpolar. Typically, most organic compounds are not soluble in water because they contain no polar groups that can hydrogen bond.

When a compound burns in the presence of oxygen, the reaction is called *combustion*. This is the reaction that occurs when the methane gas in a Bunsen burner, a gas range, or heater is ignited. The products of combustion are carbon dioxide (CO_2), and water (H_2O).

$$CH_4(g) + 2\,O_2(g) \xrightarrow{\text{heat}} CO_2(g) + 2\,H_2O(g)$$
Methane

Alkenes and alkynes react easily with halogens such as bromine (Br_2). In a reaction known as *addition*, the unsaturated hydrocarbons add bromine atoms to each carbon atom in the double or triple bond. Evidence for the addition reaction is the disappearance of the red color of the bromine solution as the double or triple bond adds the bromine atoms.

Ethene (ethylene) + Br_2 (red) → 1,2-Dibromoethane (colorless)

Ethyne (acetylene) + 2 Br_2 (red) → 1,1,2,2-Tetrabromoethane (colorless)

254

SOME PROPERTIES OF HYDROCARBONS

In a reaction that requires the presence of light or heat, alkanes can react with bromine, but quite slowly. In a *substitution* reaction, an atom of bromine replaces a hydrogen atom. The slow disappearance of the red bromine color indicates that an alkane is present. The pungent odor of HBr may also be noticed. However, if bromine is mixed with an alkane in a dark corner, or drawer, the red color will persist.

$$\underset{\text{Ethane}}{\text{H}_3\text{C}-\text{CH}_3} + \underset{(\text{red})}{\text{Br}_2} \xrightarrow{\text{heat or light}} \underset{\text{Bromoethane}}{\text{H}_3\text{C}-\text{CH}_2\text{Br}} + \text{HBr} \ (\text{colorless})$$

Although benzene appears to be unsaturated, it does not react with bromine. It behaves like an alkane rather than an alkene. In benzene, the arrangement of the electrons in the cyclic six-carbon structure provides a stable compound that does not react readily.

$$\bigcirc + \text{Br}_2 \longrightarrow \text{No change}$$

Another test for unsaturation uses potassium permanganate ($KMnO_4$) to determine the presence of an unsaturated compound. In the Baeyer test, an alkene or alkyne undergoes an oxidation that rapidly changes the purple color of the potassium permanganate ($KMnO_4$) to the brown color of manganese dioxide (MnO_2).

$$\underset{(\text{purple})}{\text{H}_2\text{C}=\text{CH}_2} + KMnO_4 \longrightarrow \underset{(\text{brown})}{\text{HOCH}_2-\text{CH}_2\text{OH}} + MnO_2$$

LABORATORY ACTIVITIES

BE SURE YOU ARE WEARING YOUR SAFETY GOGGLES!

A. MODELS OF ALKENES AND ALKYNES

Use the carbon and hydrogen atoms in the organic model kit to construct models of ethene (ethylene), propene, cyclobutene, cis-2-butene, and acetylene. Use the springs to form the double and triple bonds. A *cis* isomer has the carbon groups attached on the same side of the double bond; a *trans* isomer has the groups attached on opposite sides. Draw their condensed structural formulas.

The hydrocarbons used in this experiment are flammable. No burners are to be used in the laboratory during these procedures.

EXPERIMENT 29

B. SOLUBILITY OF HYDROCARBONS IN WATER

Place 3-4 mL of water in each of three test tubes. Add 5 drops of a hydrocarbon and stir. Observe if the hydrocarbon mixes, or if two layers forms. Test the solubility of cyclohexane, cyclohexene, and toluene.

look to see which layer is org, which is H₂O
if dispper substance is sol

C. COMBUSTION (This may be an instructor demonstration.)

C.1 **WORKING IN THE HOOD**, place 10 drops of cyclohexane on an evaporating dish. **CAREFULLY** ignite it using a long match. Repeat the combustion with cyclohexene, and toluene. Record your observations for each.

C.2 Write the equations for the combustion reactions.

D. REACTIONS WITH BROMINE (Optional)   *Red (diluted solution)*

In this experiment, a bromine solution will be mixed with cyclohexane, cyclohexene, and toluene. A reaction occurs when bromine (Br_2) adds to the unsaturated site causing the red color of the bromine to disappear quickly. With saturated compounds, or aromatic compounds, the red color of bromine will persist for a longer time, fading slowly in the presence of light.

add 1 drop @ a time

colorless — +hybromide

WORK IN THE HOOD. The fumes of Br_2 are toxic. They irritate the throat and sinuses causing painful burns. Your instructor may wish to demonstrate this portion of the experiment. Discard test tube contents in designated containers or drains.

D.1 Place 15-20 drops of each hydrocarbon in separate, dry test tubes. Carefully add 4 drops of the bromine solution to each. Observe the rapid or slow fading of the red bromine color.

D.2 From your test results, identify each compound as saturated or unsaturated.

D.3 Draw the structural formula of the reactant.

D.4 Write the structural formula of the products if a reaction occurred. If no reaction occurs, write N.R.

E. REACTIONS WITH POTASSIUM PERMANGANATE

Place 10 drops of cyclohexane in a test tube. Add 2 drops of 0.1 M $KMnO_4$ solution. A change in color from purple to brown in 30-60 seconds indicates an unsaturated compound. Repeat with cycloehexene and toluene. Record your observations.

F. REACTIONS OF A VEGETABLE OIL

Perform the solubility, combustion, bromine and potassium permanganate tests with a few drops of a vegetable oil. Record your results. Identify the type of hydrocarbon in the oil.

NAME_____ SECTION_____ DATE_____

EXPERIMENT 29
SOME PROPERTIES OF HYDROCARBONS
LABORATORY REPORT

A. MODELS OF SOME ALKENES AND AKLYNES

Name	Condensed Structural Formula
Ethene	_____
Propene	_____
cyclobutene	_____
cis-2-butene	_____
acetylene	_____

B. SOLUBILITY OF HYDROCARBONS IN WATER

Hydrocarbon	Observations
cyclohexane	appears to be more soluble than other hydrocarbons
cyclohexene	
toluene	Clear bubble under the top layer, not as soluble when tube is tapped you see oily residue

EXPERIMENT 29

C. COMBUSTION

C.1

Hydrocarbon	Observations
cyclohexane	
cyclohexene	
toluene	

C.2 Complete and balance the following equations:

Combustion of cyclohexane

$$\underline{}C_6H_{14} + \underline{}O_2 \xrightarrow{heat} \underline{}CO_2 + \underline{}H_2O$$

Combustion of cyclohexene

$$\underline{}C_6H_{12} + \underline{}O_2 \xrightarrow{heat} \underline{}CO_2 + \underline{}H_2O$$

Combustion of toluene

$$\underline{}C_7H_8 + \underline{}O_2 \xrightarrow{heat} \underline{}CO_2 + \underline{}H_2O$$

D. REACTIONS WITH BROMINE

D.1

Hydrocarbon	Observations
cyclohexane	define
cyclohexene	there was no color
toluene	no rxns

NAME_____ SECTION_____ DATE_____

D.2-D.4

Hydrocarbon	Saturated or unsaturated	Structural formula	Product with bromine
cyclohexane	_____	_____	_____
cyclohexene	_____	_____	_____
toluene	_____	_____	_____

E. REACTIONS WITH POTASSIUM PERMANGANATE

Hydrocarbon	Observations
cyclohexane	no rxn
cyclohexene	turned brown
toluene	no rxn

F. REACTIONS WITH VEGETABLE OIL

Test	Observations
Solubility	insoluble
Combustion	
Bromine	rxn — cloudy white. Purple went away
$KMnO_4$	no reaction still purple

From your observations of the reactions of a vegetable oil, describe the type of hydrocarbon.

QUESTIONS AND PROBLEMS

1. Write the names of the following compounds:

   a. $CH_3CH_2CH=CH_2$ _____

   b. $CH_2=CHCHCH_2CH_3$ with CH_3 substituent _____

   c. 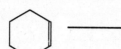 _____

   d. $CH_3C\equiv CH$ _____

2. Draw the structure of each compound:

   a. 2-chloro-2-butene b. benzene c. 1,3-dibromocyclohexene

   d. cis-2,3-dichloro-2-butene e. p-dichlorobenzene

3. Write the products of the following reactions (if no reaction, write N.R.):

   a. $CH_3CH=CHCH_3 + Br_2 \longrightarrow$

   b. $Br_2 +$ $\xrightarrow{\text{light}}$

   c. $C_3H_8 + O_2 \xrightarrow{\text{heat}}$

   d. [benzene ring] $+ Cl_2 \xrightarrow{\text{dark}}$

   e. $CH_2=CH_2 + O_2 \xrightarrow{\text{heat}}$

EXPERIMENT 30
PROPERTIES OF ALCOHOLS
ALDEHYDES AND KETONES

GOALS

1. Observe the solubility of alcohols, aldehydes and ketones in water.
2. Classify alcohols as primary, secondary, or tertiary.
3. Observe the oxidation of alcohols and aldehydes.
4. Write the structural formulas of the oxidation products of alcohols and aldehydes.

MATERIALS NEEDED

organic model kits
test tubes and test tube rack
alcohols: ethyl alcohol, 1-propanol, 2-propanol, t-butyl alcohol (2-methyl-2-propanol), cyclohexanol
hot water bath
Benedict's reagent
aldehydes: acetaldehyde, propanal, benzaldehyde, cinnamaldehyde, vanillin
ketones: acetone (2-propanone), camphor
5% glucose solution
2% $K_2Cr_2O_7/H_2SO_4$ solution
chemistry hand book

CONCEPTS TO REVIEW

alcohols
classification of alcohols
solubility of alcohols in water
aldehydes
ketones
oxidation of alcohols and aldehydes

BACKGROUND DISCUSSION

Classification of Alcohols

Alcohols are organic compounds containing at least one hydroxyl group(-OH). They can be classified by the number of carbon atoms held by the carbon atom attached to the -OH group. In a *primary (1°)* alcohol, the carbon atom attached to the -OH group is bonded to one other carbon atom. Methyl alcohol is included in this primary group. A *secondary (2°)* alcohol has two carbon atoms attached to the carbon holding the -OH group. In a *tertiary (3°)* alcohol, there are three carbon atoms attached to the carbon with the -OH group.

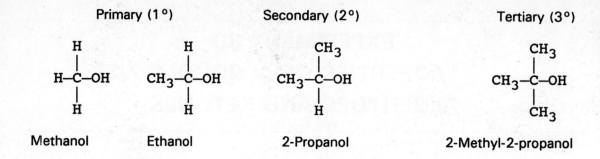

Solubility of Alcohols

The alcohols containing one to four carbon atoms are soluble in water due to the hydrogen bonding of the polar hydroxyl group. Alcohols with longer nonpolar carbon chains are not soluble in water.

$$CH_3-O-H$$
$$:$$
$$O-H$$
$$|$$
$$H$$

Hydrogen bonding of an alcohol with water

Aldehydes and ketones contain the carbonyl functional group. In aldehydes, the carbonyl group occurs at the end of the carbon chain. In ketones, the carbonyl group is attached to two of the carbon atoms within the chain.

$-\overset{O}{\underset{\|}{C}}-$	$CH_3\overset{O}{\underset{\|}{C}}H$	$CH_3CH_2\overset{O}{\underset{\|}{C}}H$	$CH_3\overset{O}{\underset{\|}{C}}CH_3$	$-\overset{O}{\underset{\|}{C}}H$
carbonyl group	Acetaldehyde	Propionaldehyde (Propanal)	Acetone (2-Propanone)	Benzaldehyde

Oxidation of Alcohols

Primary and secondary alcohols can be oxidized to aldehydes and ketones. In an oxidation, one hydrogen is removed from the hydroxyl group of the alcohol, and one hydrogen is removed from the carbon attached to the -OH group. In this experiment, potassium dichromate, $K_2Cr_2O_7$, is used as the oxidizing agent. Initially, the potassium dichromate is orange in color. When oxidation occurs, the dichromate is converted to chromium (III) ion, Cr^{3+}, which is green. Typically, the aldehyde obtained from the oxidation of the primary alcohol will be oxidized further by excess $K_2Cr_2O_7$ to give a carboxylic acid.

$$CH_3-CH_2-OH + K_2Cr_2O_7 \xrightarrow{H^+} CH_3-\overset{O}{\underset{\|}{C}}-H + 2\ Cr^{3+}$$

primary (1°) alcohol   (orange)   aldehyde   (green)

PROPERTIES OF ALCOHOLS, ALDEHYDES AND KETONES

Secondary alcohols oxidize to give ketones.

$$CH_3-\underset{\underset{\text{secondary (2°)}}{|}}{\overset{OH}{C}H}-CH_3 + K_2Cr_2O_7 \xrightarrow{H^+} CH_3-\overset{O}{\underset{|}{C}}-CH_3 + 2\ Cr^{3+}$$

secondary (2°) alcohol   (orange)   ketone   (green)

Tertiary alcohols do not undergo oxidation; there are no hydrogen atoms on the carbon atom attached to the hydroxyl group.

$$CH_3-\underset{\underset{CH_3}{|}}{\overset{\overset{CH_3}{|}}{C}}-OH + K_2Cr_2O_7 \xrightarrow{H^+} \text{No reaction}$$

tertiary (3°) alcohol   (orange)

Oxidation of Aldehydes

Aldehydes are easily oxidized to give carboxylic acids. This reaction distinguishes aldehydes from ketones which cannot undergo further oxidation. In this experiment, we will use Benedict's reagent which contains a mild oxidizing agent, cupric ion, Cu^{2+}, in a basic solution. When the aldehyde group is oxidized, the cupric ion is reduced to cuprous ion (Cu^+) which forms a reddish-orange precipitate, Cu_2O.

$$CH_3\overset{O}{\overset{\|}{C}}H + 2\ Cu^{2+} \xrightarrow{\text{oxidation}} CH_3\overset{O}{\overset{\|}{C}}OH + Cu_2O(s)$$

aldehyde   (blue)   carboxylic acid   (red-orange)

$$CH_3\overset{O}{\overset{\|}{C}}CH_3 + 2\ Cu^{2+} \xrightarrow{\text{oxidation}} \text{No reaction}$$

(blue)

    Glucose is an example of a carbohydrate that contains an aldehyde group. The oxidation of glucose with Benedict's reagent is a typical clinical test for detecting glucose in the urine.

EXPERIMENT 30

$$\begin{array}{c}
O \\
\parallel \\
C-H \\
| \\
H-C-OH \\
| \\
HO-C-H \\
| \\
H-C-OH \\
| \\
H-C-OH \\
| \\
CH_2OH
\end{array}$$

D-Glucose

LABORATORY ACTIVITIES

A. MODELS OF ALCOHOLS, ALDEHYDES, AND KETONES

Obtain an organic model kit and construct a model of 1-propanol. Convert this model to 2-propanol, propanal (propionaldehyde), and 2-propanone (acetone). Use the spring connectors for the double bonds. Write the condensed structural formula of each.

B. SOLUBILITY IN WATER

Place 3-4 mL of water in a test tube. Add 5 drops of the compound to test. Stir and observe the resulting solution. If the substance is soluble in water, you will see a clear solution with no separate layers. If it is insoluble, a cloudy or separate layer will form. If the substance is soluble, add another 5 drops and observe the solution again. Record your observations. Indicate whether each substance was soluble or insoluble. Give a brief reason. Test the solubility of the following :

> ethyl alcohol, 1-propanol, cyclohexanol,
> acetaldehyde, propanal, and 2-propanone

C. OXIDATION OF ALCOHOLS

C.1   Draw the structural formulas of the following alcohols:

> ethyl alcohol, 1-propanol, 2-propanol,
> t-butyl alcohol (2-methyl-2-propanol)

C.2   Identify each alcohol as a primary (1°), secondary (2°), or tertiary (3°) alcohol.

C.3   Carefully add 1 mL (20 drops) of each alcohol to separate test tubes. *Cautiously* add 10 drops of 2% $K_2Cr_2O_7/H_2SO_4$ solution to each.

PROPERTIES OF ALCOHOLS, ALDEHYDES AND KETONES

 CAUTION: THE POTASSIUM DICHROMATE SOLUTION CONTAINS CONCENTRATED H_2SO_4 WHICH CAN BURN THE SKIN.

If a test tube becomes hot, place it in a beaker of ice-cold water. Look for a color change from orange to green. If the color remains orange, no reaction has occurred. If the orange turns to green, oxidation of the alcohol has taken place. Record your observations.

C.4　Write the structural formulas for the oxidation products. Where there is no change in color, write no reaction, N.R.

D.　PHYSICAL PROPERTIES OF ALDEHYDES AND KETONES

D.1　Using a chemistry handbook or the Merck Index, write the structural formula of camphor, cinnamaldehyde, benzaldehyde and vanillin.

D.2　Circle the functional group in each structure and name the group.

D.3　Record the melting and boiling point of each.

D.4　Carefully note the odor of each and record.

D.5　Give some commercial or medicinal uses for the compounds.

E.　OXIDATION OF ALDEHYDES AND KETONES

E.1　Write the structural formulas of acetaldehyde, propanal, acetone, and glucose.

E.2　Place 1 mL of each substance in separate test tubes and label. Add 2 mL of Benedict's reagent to each test tube. Place the test tubes in a hot water bath for 10 minutes. A change in color from blue to various shades of green to a reddish color of Cu_2O indicates that oxidation has occurred. Record your observations.

 CAUTION: ACETONE IS FLAMMABLE.

E.3　Write the structural formulas of the oxidation products. If no reaction occurs, write N.R.

NAME_____ SECTION_____ DATE_____

EXPERIMENT 30
PROPERTIES OF ALCOHOLS
ALDEHYDES AND KETONES
LABORATORY REPORT

A. MODELS OF ALCOHOLS, ALDEHYDES, AND KETONES

Name	Condensed Structural Formula
1-Propanol	
2-Propanol	
Propanal	
2-Propanone	

B. SOLUBILITY IN WATER

Compound	Observations
Ethyl alcohol	
1-Propanol	
Cyclohexanol	
Acetaldehyde	
Propanal	
2-Propanone	

EXPERIMENT 30

Which of the compounds tested in B were soluble in water?
_____

Give a brief reason for their solubility.

C. OXIDATION OF ALCOHOLS

Compound	Structural formula (C.1)	Classification (C.2)	Color change (C.3)	Oxidation product (C.4)
Ethyl alcohol				
1-Propanol				
2-Propanol				
t-Butyl alcohol				

Which compound(s) did not oxidize? Why?

NAME_____ SECTION_____ DATE_____

D. PHYSICAL PROPERTIES OF ALDEHYDES AND KETONES

Compound	Structural formula (D.1, D.2)	Melting and boiling points (D.3)	Odor (D.4)
Camphor			
Cinnamaldehyde			
Benzaldehyde			
Vanillin			

EXPERIMENT 30

D.5 What are some commercial or medicinal uses associated with these aldehydes and ketones?

Camphor

Cinnamaldehyde

Benzaldehyde

Vanillin

E. OXIDATION OF ALDEHYDES AND KETONES

Compound	Structural formula (E.1)	Color change (E.2)	Oxidation product (E.3)
Acetaldehyde			
Propanal			
Acetone			
Glucose			

NAME_____SECTION_____DATE_____

QUESTIONS AND PROBLEMS

1. Write the structures and classifications of the following alcohols:

   a. 2-Pentanol

   c. 2-Methyl-2-butanol

   b. 3-Methylcyclopentanol

   d. 1-Propanol

2. Write the names of the following compounds:

   a. $CH_3-CH_2-CH_2-CH_2-\overset{\overset{O}{\|}}{C}H$

   c. cyclopentanone (structure shown)

   _____   _____

   b. $CH_3-\overset{\overset{O}{\|}}{C}-CH_2-\overset{\overset{CH_3}{|}}{C}H-CH_3$

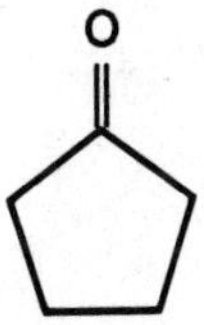

   d. 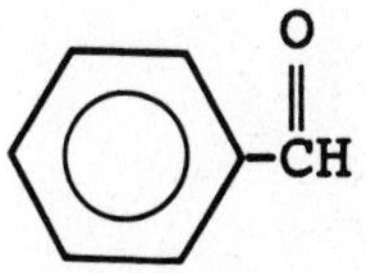 benzaldehyde (structure shown)

   _____   _____

EXPERIMENT 30

3. Write the product of the following reactions (if no reaction, write N.R.):

   a. $CH_3-CH_2-CH_2OH \xrightarrow{\text{oxidation}}$

   b. $CH_3-\underset{\underset{\displaystyle CH_3}{|}}{CH}_2-\overset{\overset{\displaystyle O}{\|}}{C}OH \xrightarrow{\text{oxidation}}$

   c. $C_6H_5-CHO \xrightarrow{\text{oxidation}}$

   d. $CH_3CH_2-\underset{\underset{\displaystyle OH}{|}}{CH}-CH_3 \xrightarrow{\text{oxidation}}$

   e. cyclohexanol $\xrightarrow{\text{oxidation}}$

EXPERIMENT 31
CARBOXYLIC ACIDS AND ESTERS

GOALS

1. Write the structural formulas of carboxylic acid and esters.
2. Determine the solubility in water of carboxylic acids and their salts.
3. Observe the formation of esters.
4. Identify the fragrance of an ester.
5. Observe the saponification of an ester.

MATERIALS NEEDED

test tubes and test tube rack
carboxylic acids: acetic acid, propionic acid, tartaric acid, benzoic acid, salicylic acid
red and blue litmus paper
alcohols: methanol, pentyl alcohol, ethanol, benzyl alcohol, octanol
glacial acetic acid
hot water bath
concentrated H_3PO_4
10% NaOH
10% HCl
methyl salicylate

DISCUSSION OF EXPERIMENT

Carboxylic acids contain the carboxyl group. Examples such as acetic acid which we know as vinegar are shows below:

$$-\overset{\overset{O}{\|}}{C}-OH \quad \text{Carboxyl group}$$

$$CH_3\overset{\overset{O}{\|}}{C}OH \qquad CH_3CH_2\overset{\overset{O}{\|}}{C}OH \qquad C_6H_5-\overset{\overset{O}{\|}}{C}OH \qquad HO\overset{\overset{O}{\|}}{C}-\overset{OH}{\underset{}{C}H}-\overset{OH}{\underset{}{C}H}-\overset{\overset{O}{\|}}{C}OH$$

Acetic acid   Propionic acid   Benzoic acid   Tartaric acid

EXPERIMENT 31

Ionization of Carboxylic Acids in Water

The polarity of the carboxyl group makes low molecular weight acids, and diacids with two or more carboxyl groups soluble in water. They act as weak acids in aqueous solution when a small percent of the carboxyl groups dissociate.

$$CH_3-\underset{\underset{O}{\parallel}}{C}-OH \xrightarrow{H_2O} CH_3-\underset{\underset{O}{\parallel}}{C}-O^- + H^+$$

Acetic acid → Acetate ion (a carboxylate ion)

Neutralization of Carboxylic Acids

Carboxylic acids are neutralized by bases such as sodium hydroxide to form carboxylate salts and water. These salts are usually soluble in water.

$$CH_3-\underset{\underset{O}{\parallel}}{C}-OH + NaOH \longrightarrow CH_3-\underset{\underset{O}{\parallel}}{C}O^-\ Na^+ + H_2O$$

Acetic acid → Sodium acetate

Esters

While carboxylic acids may have tart or unpleasant odors, many esters have pleasant, fragrant odors.

$$CH_3-\underset{\underset{O}{\parallel}}{C}-O(CH_2)_7CH_3 \qquad CH_3-\underset{\underset{O}{\parallel}}{C}-OCH_2(CH_2)_3CH_3$$

Octyl acetate (oranges)   Pentyl acetate (pears)   Methyl salicylate (wintergreen)

Esterification and Hydrolysis

An ester can form when a carboxylic acid and an alcohol react in the presence of an acid catalyst such as H_2SO_4. Usually, the esterification reaction requires heat.

$$CH_3-\underset{\underset{O}{\parallel}}{C}-OH + HO-CH_2CH_3 \xrightarrow[\text{heat}]{H^+} CH_3-\underset{\underset{O}{\parallel}}{C}-O-CH_2CH_3 + H_2O$$

Acetic acid   Ethanol   Ethyl acetate

The reverse reaction called *hydrolysis* occurs when an acid catalyst and water causes the decomposition of an ester to give back the carboxylic acid and alcohol.

$$\underset{\text{Methyl acetate}}{CH_3-\overset{O}{\underset{\|}{C}}-O-CH_3} + H_2O \xrightarrow{H^+} \underset{\text{Acetic acid}}{CH_3-\overset{O}{\underset{\|}{C}}-OH} + \underset{\text{Methyl alcohol}}{CH_3-OH}$$

Saponification

An ester undergoes saponification when it is hydrolyzed in the presence of a base. The products are the salt of the carboxylic acid and the alcohol.

$$\underset{\text{Ethyl acetate}}{CH_3-\overset{O}{\underset{\|}{C}}-OCH_2CH_3} + NaOH \longrightarrow \underset{\text{Sodium acetate}}{CH_3-\overset{O}{\underset{\|}{C}}O^-Na^+} + \underset{\text{Ethanol}}{CH_3CH_2-OH}$$

LABORATORY ACTIVITIES

 BE SURE YOU HAVE YOUR SAFETY GOGGLES ON!

A. SOLUBILITY OF CARBOXYLIC ACIDS AND THEIR SALTS

A.1   Write the structural formulas for acetic acid, propionic acid, tartaric acid, and benzoic acid.

A.2   Place 3-4 mL of water in four separate test tubes. Label each. Add 5 drops of the acid, if liquid or the crystals that fit on the tip of a microspatula and stir. If the acid dissolves, add 5 more drops (or another scoop). Record your observations.

A.3   Place the test tubes in a hot water bath and heat for 5 minutes. Record your observations.

A.4   Remove the test tubes and place in cold water. Test the pH of each sample by using a glass stirring rod to place a drop of the solution on the piece of pH paper. Match the color on the container and report the pH of each.

A.5   Make each solution basic by adding 10-15 drops of 10% NaOH. The solution is basic when a drop turns red litmus blue. Stir and record your observations.

A.6   Write the structural formulas of the sodium salts of each carboxylic acid.

A.7   Write the equation for the neutralization of the carboxylic acids.

EXPERIMENT 31

B. PREPARATION OF SOME ESTERS

B.1 Place the following combinations in four separate test tubes. Label each.

	Alcohol	Carboxylic acid
Ester 1	2 mL pentyl alcohol	2 mL glacial acetic acid
Ester 2	2 mL methanol	1 g salicylic acid
Ester 3	2 mL octanol	2 mL glacial acetic acid
Ester 4	2 mL benzyl alcohol	2 mL glacial acetic acid

CAUTIOUSLY add 4-5 drops of concentrated H_3PO_4. Stir the solution and place the test tube in a boiling hot water bath for about 10 minutes.

CAUTION: USE CARE IN DISPENSING GLACIAL ACETIC ACID. IT CAN CAUSE BURNS AND BLISTERS ON THE SKIN.

CAUTION: USE CARE WHEN USING CONCENTRATED PHOSPHORIC ACID.

CAUTIOUSLY note the odor of each ester. Record the fragrances you detect. If no odor is detectable, slowly add 5 mL of warm water to the test tube. Try to identify the odors of the fruit it is associated with such as pear, banana, or oil of wintergreen.

B.2 Write the structural formulas and names of the esters produced.

C. SAPONIFICATION OF AN ESTER

C.1 Place 5 mL of water in a test tube. Add 10 drops of methyl salicylate. Record the appearance and odor of the ester.

C.2 Add 2 mL of 10% NaOH. Place the test tube in a boiling hot water bath for 20-30 minutes. The top layer should disappear. Record any changes in the odor and appearance of the ester. Remove the test tube and cool in cold water.

C.3 Write the equation for the saponification reaction.

C.4 To the cooled solution, add 10% HCl until the solution is acidic (a drop turns blue litmus red). Record your observations.

NAME_____ SECTION_____ DATE_____

EXPERIMENT 31
CARBOXYLIC ACID AND ESTERS
LABORATORY REPORT

A. SOLUBILITY OF CARBOXYLIC ACIDS AND THEIR SALTS

	Acetic acid	Propionic acid	Tartaric acid	Benzoic acid
(A.1) Structural formulas				
(A.2) Solubility in cold water				
(A.3) Solubility in hot water				
(A.4) pH				
(A.5) Solubility in NaOH				
(A.6) Structural formulas of salts				

EXPERIMENT 31

A.7 Equations for neutralization

acetic acid _____

propionic acid _____

tartaric acid _____

benzoic acid _____

B. PREPARATION OF SOME ESTERS

(B.1) Alcohol	Carboxylic acid	(B.2) Odor of ester	(B.3) Structural formula

NAME_____ SECTION_____ DATE_____

C. SAPONIFICATION OF AN ESTER

C.1　Appearance and odor
　　　of methyl salicylate　　　_____

C.2　Appearance and odor
　　　after adding NaOH　　　_____

C.3　Equation for saponification

　　　_____

C.4　Observations with HCl　　_____

EXPERIMENT 32
PREPARATION OF ASPIRIN

GOALS

1. Use laboratory methods to prepare aspirin.
2. Purify the crude aspirin sample.
3. Test the prepared aspirin and aspirin products for purity.

MATERIALS NEEDED

salicylic acid
125 mL Erlenmeyer flask
5- or 10 mL graduated cylinder
thermometer
stirring rod
acetic anhydride
concentrated H_3PO_4
melting point apparatus
ice water
watch glass
Buchner funnel setup
filter paper
1% $FeCl_3$
test tubes (4)
commercial aspirin

CONCEPTS TO REVIEW

carboxylic acids
formation of esters

BACKGROUND DISCUSSION

In the 18th century, an extract of willow bark was found useful in reducing fevers and relieving pain. Today, we know that the substance in the extract was salicylic acid. By the beginning of the 19th century, it was modified by forming the ester with acetic acid to give acetylsalicylic acid or aspirin. In commercial aspirin products, a small amount of acetylsalicylic acid (typically 5 grains or 300 mg) is bound together with a starch binder and sometimes buffers to make as aspirin tablet. In the small intestine, the basic conditions cause the hydrolysis of the acetylsalicylic to give the salicylic acid which is absorbed into the blood stream.

EXPERIMENT 32

Aspirin (acetylsalicyclic acid) can be prepared from acetic acid and the hydroxyl group on salicylic acid. However, this is a slow reaction. The ester forms more rapidly when acetic anhydride is used to provide the acetyl group.

Salicylic acid   Acetic anhydride   Acetylsalicylic   Acetic
                                                            acid (aspirin)   acid

Recrystallization and Purity of Prepared Aspirin

One method used to obtain a product with less impurity is to dissolve the impure aspirin product in alcohol and allow it to recrystallize. Many of the impurities such as excess salicylic acid remain in the alcohol. The purity of the crude and recrystallized aspirin can be determined by adding ferric chloride. The iron (III) ion reacts with the phenol group of salicylic acid to a purple color. This test can be used to detect the presence of unreacted salicylic acid in impure, pure, and commercially prepared aspirin.

LABORATORY ACTIVITIES

WEAR PROTECTIVE GLASSES!

A. PREPARATION OF ASPIRIN

A.1 Weigh a 125 mL Erlenmeyer flask to the nearest 0.1 g. Prepare a hot water bath using a 400-mL beaker. Place 2.0 g of salicylic acid in a 125-mL Erlenmeyer flask. Reweigh the flask and salicylic acid. Calculate the mass of salicylic acid.

Using a 5 or 10-mL graduated cylinder, *cautiously* add 5.0 mL of acetic anhydride.

CAUTION: ACETIC ANHYDRIDE IS VERY IRRITATING TO THE MEMBRANES OF THE NOSE AND SINUS. WEAR GOGGLES.

Slowly add 10 drops of concentrated (85%) phosphoric acid. Place a thermometer in the mixture and stir slowly. The temperature will begin to rise as reaction takes place. Place the flask and thermometer in a water bath. Maintain a temperature of 50-70°C for 15-20 minutes. Be sure you continue to stir the mixture during this time. **CAUTIOUSLY** add 10 drops of water to the cooled mixture. Then add 5 mL of water. Any unreacted acetic anhydride will decompose and emit acetic acid vapors. Remove the flask and cool the mixture in an ice-water bath. Add 20 mL of water.

KEEP YOUR FACE AWAY FROM THE TOP OF THE FLASK. THE DECOMPOSITION OF ACETIC ANHYDRIDE CAN BE VERY REACTIVE. ACETIC ACID VAPORS ARE EMITTED THAT ARE VERY IRRITATING.

If no crystals appear when the contents have cooled to room temperature, scratch the sides with a stirring rod. Crystals of aspirin should form.

Collect the aspirin by pouring the water and crystals into a Buchner funnel using suction filtration. (See Figure 32.1) Remove any remaining crystals by rinsing the flask with 5 mL portions of ice water. Draw air through the funnel to remove as much water as possible from the crystals.

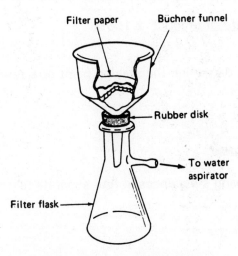

Figure 32.1 Apparatus for suction filtration.

A.2 Weigh a watch glass to the nearest 0.1 g. Transfer the impure aspirin crystals to the watch glass and reweigh. This gives the mass of the crude product.

A.3 Calculate the percentage yield of crude aspirin product. The theoretical yield of aspirin from 2.0 g salicylic acid would be 2.6 g.

$$\text{percent yield} = \frac{\text{mass of product}}{2.6 \text{ g}} \times 100\%$$

A.4 Using a melting point apparatus, determine the melting point of the crude product. Record.

A.5 Obtain a chemistry handbook and look up the actual melting point of aspirin (acetylsalicylic acid).

NOTE: *If time permits, set aside a few crystals of the impure aspirin, and proceed to the section on recrystallization of the crude aspirin product. If not, proceed to Section C to determine the purity of aspirin.*

EXPERIMENT 32

B. RECRYSTALLIZATION OF CRUDE ASPIRIN PRODUCT (OPTIONAL)

The major impurity in the aspirin product is unreacted salicylic acid. To purify your crude aspirin product, place the impure aspirin in a 125 mL Erlenmeyer flask. Add 15 mL of ethanol. If crystals remain, add 30 mL warm water and warm the flask on a hot plate until all the crystals dissolve. Allow the solution to cool. You may cool further by placing the beaker in ice. Crystals of purified aspirin should form. Collect the purified product with a Buchner funnel as you did before. Draw air through to dry the crystals.

B.1 Weigh a watch glass to the nearest 0.1 g. Transfer the dry aspirin crystals to the watch glass and reweigh.

B.2 Calculate the percent yield of the purified aspirin product. (This may be low.)

$$\text{percent yield} = \frac{\text{mass of product}}{2.6 \text{ g}} \times 100\%$$

B.4 Using a melting point apparatus, determine the melting point of a few crystals of the purified aspirin.

C. DETERMINING THE PURITY OF ASPIRIN

C.1 Place a few crystals of the following substances in four separate test tubes.

(1) salicylic acid
(2) crude product
(3) purified product
(4) commercial aspirin

Add 3 mL of water and 2 drops of 1% iron (III) chloride ($FeCl_3$) solution to each. Record the color. Explain your observations.

Optional: Place your aspirin product in a small vial. Attach a label with your name(s), mass of product, percent yield (which may be low), and melting point. Turn in to your instructor.

EXPERIMENT 32
PREPARATION OF ASPIRIN
LABORATORY REPORT

A. PREPARATION OF ASPIRIN

A.1 Mass of flask and salicylic acid and flask _____ g

Mass of flask _____ g

Mass of salicylic acid _____ g

A.2 Mass of watch glass and impure aspirin _____ g

Mass of watch glass _____ g

Mass of impure aspirin _____ g

A.3 Percent yield of impure aspirin _____ %
Show calculations:

A.4 Melting point of impure aspirin _____ °C

A.5 Actual melting point of aspirin _____ °C

EXPERIMENT 32

B. RECRYSTALLIZATION OF CRUDE ASPIRIN PRODUCT (OPTIONAL)

B.1 Mass of watch glass and pure aspirin _____ g

      Mass of watch glass _____ g

      Mass of pure aspirin _____ g

B.2 Percent yield of pure aspirin _____ %
Show calculations:

B.3 Melting point of pure aspirin _____ °C

C. DETERMINING THE PURITY OF ASPIRIN

	Color with $FeCl_3$	Pure/impure?
(1) Salicylic acid	_____	_____
(2) Impure aspirin	_____	_____
(3) Pure aspirin	_____	_____
(4) Commercial aspirin	_____	_____

NAME_____SECTION_____DATE_____

QUESTIONS AND PROBLEMS

1. Write the structural formula for aspirin. Label the ester group and the carboxylic acid group.

2. What chemical change does aspirin undergo before it is absorbed into the bloodstream?

3. Which of the samples tested in part C would be considered impure? Why?

4. Aspirin that has been stored for a long time may give a vinegar-like odor and give a purple color with $FeCl_3$. What reaction would cause this to happen?

EXPERIMENT 33
AMINES AND AMIDES

GOALS

1. Write the structural formulas and/or names of amines.
2. Classify amines as primary, secondary or tertiary.
3. Observe some physical properties of amines.
4. Write an equation for the formation of an amine salt.
5. Write equations for amidation and hydrolysis reactions.

MATERIALS NEEDED

organic model kits
test tubes
diethylamine
acetamide
methyl benzoate
con. ammonia
10% HCl
10% NaOH
impure acetanilide

Buchner funnel
short stem funnel
filter paper
pH and litmus paper
250 mL beaker
250 mL Erlenmeyer flask
melting point apparatus
watch glass

CONCEPTS TO REVIEW

amines
naming amines
amides
solubility and pH of amines
amidation
hydrolysis of amines

DISCUSSION OF EXPERIMENT

Amines are derivatives of ammonia. In an amine, alkyl (or aromatic) groups replace one or more hydrogens of ammonia to give primary, secondary or tertiary amines.

NH_3	CH_3-NH_2	$CH_3-NH-CH_3$	$CH_3-\underset{\underset{CH_3}{\mid}}{N}-CH_3$
Ammonia	Methylamine (primary)	Dimethylamine (secondary)	Trimethylamine (tertiary)

EXPERIMENT 33

Amines in water

In water, ammonia and amines act as weak bases because the unshared pair of electrons on the nitrogen atom attracts protons. An ammonium ion or alkyl ammonium ions is produced along with a hydroxide ion. The OH⁻ ions give a basic test with red litmus paper.

$$NH_3 + H_2O \longrightarrow NH_4^+ + OH^-$$

Ammonia → Ammonium ion

$$CH_3NH_2 + H_2O \longrightarrow CH_3NH_3^+ + OH^-$$

Methylamine → Methylammonium ion

Neutralization of Amines

Amines react with acids to form the amine salt. These salts are much more soluble in water than the corresponding amines.

$$CH_3NH_2 + HCl \longrightarrow CH_3NH_3^+ Cl^-$$

Methylamine → Methylammonium chloride (amine salt)

Amides

Amides contain the amide group as seen in the examples below:

$$-\overset{O}{\underset{}{\overset{\|}{C}}}-\overset{|}{N}- \qquad CH_3-\overset{O}{\overset{\|}{C}}-NH_2 \qquad C_6H_5-\overset{O}{\overset{\|}{C}}-NH_2 \qquad CH_3\overset{O}{\overset{\|}{C}}-NHCH_3$$

Amide · · · · · Acetamide · · · · · Benzamide · · · · · N-Methylacetamide

In a reaction called amidation, an amide forms when a carboxylic acid is heated with ammonia or an alkyl (aromatic) amine.

$$CH_3-\overset{O}{\overset{\|}{C}}-OH + CH_3-NH_2 \xrightarrow{\text{heat}} CH_3-\overset{O}{\overset{\|}{C}}-NH-CH_3$$

Acetic acid · · · · · Methylamine · · · · · N-Methylacetamide

Hydrolysis of Amides

An amide is hydrolyzed by either an acid or a base. Acid hydrolysis produces the carboxylic acid and ammonium salt. Base hydrolysis gives the carboxylate salt and ammonia which can be used to detect the hydrolysis reaction.

$$CH_3-\underset{\underset{O}{\|}}{C}-NH_2 + HCl \xrightarrow{heat} CH_3-\underset{\underset{O}{\|}}{C}OH + NH_4Cl$$

$$CH_3-\underset{\underset{O}{\|}}{C}-NH_2 + NaOH \xrightarrow{heat} CH_3-\underset{\underset{O}{\|}}{C}O^-Na^+ + NH_3(g)$$

LABORATORY ACTIVITIES

WEAR SAFETY GOGGLES!

A. STRUCTURE AND CLASSIFICATION OF AMINES

A.1 Use a model kit to prepare or observe models of the following:

ammonia, methylamine, dimethylamine, trimethylamine

Write the condensed structure of each model in the laboratory report.

A.2 Classify each amine as primary (1°), secondary (2°), or tertiary (3°).

B. IONIZATION OF AMINES IN WATER

WORK IN THE HOOD. THE VAPORS OF AMINES ARE IRRITATING TO NOSE AND SINUS.

B.1 Place 5 mL of water in a test tube. Add 10 drops of diethylamine. *Cautiously* note its odor. Record.

B.2 Stir with a glass stirring rod and observe the solubility of diethylamine in water.

B.3 Touch the glass stirring rod in the amine to pH paper. Record your observations.

B.4 Write the equation for the ionization of diethylamine in water.

B.5 Add 10% HCl dropwise to the amine solution until the solution is acidic with litmus paper. Record any changes in the odor and/or solubility of the amine.

B.6 Write the neutralization equation for the reaction of the amine with HCl.

EXPERIMENT 33

C. FORMATION OF AN AMIDE *This may be demonstrated by your instructor.*

The amide, benzamide, may be produced by combining methyl benzoate and concentrated ammonia. **USE CAREFULLY!**

$$C_6H_5\text{-COOCH}_3 + NH_3 \longrightarrow C_6H_5\text{-CONH}_2 + CH_3OH$$

    Methyl benzoate   Ammonia   Benzamide   Methanol

Place 10 mL of methyl benzoate and 30 mL of concentrated ammonia in a 250-mL Erlenmeyer flask. Stopper and set in the hood. Do not disturb. At the next laboratory, look for the formation of crystals of benzamide. If crystals are present, filter and discard the filtrate.

CAUTION: THE MIXTURE STILL CONTAINS CONCENTRATED AMMONIA.

Allow the crystals to dry. Take a melting point and compare in to the actual melting point of benzamide. Record.

D. HYDROLYSIS OF AN AMIDE

D.1 Place a small amount of acetamide crystals in each of two test tubes. **CAUTIOUSLY** note the odor.

D.2 Add 2 mL of 10% HCl to the first test tube, and 2 mL of 10 % NaOH to the second. Place the test tubes in a boiling water bath and heat gently. **CAUTIOUSLY** note the odor of any gas coming from each mixture. Record.

D.3 Hold a piece of moist litmus paper over the mouth of each test tube. Record any change in the color of the litmus paper.

D.4 Write the equations for the acid and base hydrolysis of acetamide.

E. PURIFICATION OF IMPURE ACETANILIDE

The amide acetanilide shows a tenfold increase in solubility from 25°C to 100°C. This difference in solubility can be used to isolate and purity acetanilide from an impure sample.

E.1 Weigh a 250-mL beaker. Add about 2 g of impure acetanilide. Reweigh the beaker and contents. Record. Calculate the mass of the impure sample.

Purification procedure

Add 50 mL of water to the beaker and heat the mixture until no more solid material appears to dissolve. While the mixture is heating, heat another beaker of water for use in the filtration.

AMINES AND AMIDES

Place a short-stem funnel fitted with filter paper in an iron ring. When you are ready to filter your impure sample, pour some of the hot water through the funnel to warm the glass. Discard the water.

Filter the warm sample of impure acetanilide into a clean beaker or flask. Rinse the funnel with hot water before crystals form on it. Place the beaker or flask containing the warm filtrate in an ice bath. As the filtrate cools, crystals of acetanilide should form.

E.2   Weigh a watch glass. Collect the crystals of acetanilide using the Buchner funnel and suction filtration apparatus. Place the crystals on the weighed watch glass and weigh. Determine the mass of the product and record.

E.3   Calculate the percent yield of the pure product.

$$\% \text{ yield} = \frac{\text{mass of purified product}}{\text{mass of impure sample}} \times 100\%$$

E.4   Use a melting point apparatus to determine the melting point of the purified acetanilide. Use a chemistry handbook to determine the actual melting point. Record.

Place the purified product in a small vial, stopper, label with your name, % yield and melting point of the product. Turn in to your instructor.

NAME_____ SECTION_____ DATE_____

EXPERIMENT 33
AMINES AND AMIDES
LABORATORY REPORT

A. STRUCTURE AND CLASSIFICATION OF AMINES

Compound	Structure (A.1)	Classification (A.2)
Ammonia		
Methylamine		
Dimethylamine		
Trimethylamine		

B. IONIZATION OF AMINES IN WATER

B.1 Odor of diethylamine _____

B.2 Solubility _____

B.3 pH _____

   Acidic or basic? _____

B.4 Equation for ionization of diethylamine in water

_____

B.5 Observations after adding HCl

   Odor _____

   Solubility _____

B.6 Equation for neutralization of diethylamine with HCl

_____

EXPERIMENT 33

C. FORMATION OF AN AMIDE

Melting point of product _____ °C

Handbook value for the
melting point of benzamide _____ °C

D. HYDROLYSIS OF AN AMIDE

D.1 Odor of acetamide _____

D.2 Odor after adding acid _____

Odor after adding base _____

D.3 Color of litmus paper in
acid hydrolysis _____

base hydrolysis _____

D.4 Equations for the hydrolysis of benzamide

with HCl _____

with NaOH _____

E. PURIFICATION OF IMPURE ACETANILIDE

E.1 Mass of beaker and impure sample _____ g

Mass of beaker _____ g

Mass of impure sample _____ g

E.2 Mass of watch glass + pure product _____ g

Mass of watch glass _____ g

Mass of pure product _____ g

E.3 Percent yield _____ g
Show calculations:

E.4 Melting point of product _____ °C

Handbook value for melting
point of acetanilide _____ °C

EXPERIMENT 34
CARBOHYDRATES

GOALS

1. Observe some physical and chemical characteristics of typical carbohydrates.
2. Use chemical tests to differentiate between monosaccharides, disaccharides, and polysaccharides.
3. Identify an unknown carbohydrate.

MATERIALS NEEDED

1% solutions: glucose, fructose, sucrose, lactose, maltose, starch
test tubes
boiling water bath
reagents: Benedict's, iodine, Seliwanoff's
Baker's yeast
6M HCl
6M NaOH
fermentation tubes (or test tubes with smaller tubes to fit inside)
sugars (refined, brown,"natural", powdered), honey
syrups (corn, maple, fruit)
foods: cereals, pasta, bread, crackers, potato

CONCEPTS TO REVIEW

carbohydrates
aldohexoses
ketohexoses
Fischer projection
Haworth structure
reducing sugars

BACKGROUND DISCUSSION

Carbohydrates are the major compounds in our diet that provide energy to run chemical reactions in the body and make us move. Foods high in carbohydrate include potatoes, bread, pasta, and rice. While we use carbohydrates to meet our energy requirements, we don't need too much. Excess carbohydrate intake is converted to fat, and makes us overweight.

EXPERIMENT 34

Chemically, carbohydrates are polyhydroxy compounds that have a carbonyl group. They can be aldehydes or ketones. Glucose is a typical monosaccharide while maltose is a typical disaccharide. Glucose exists almost entirely in the closed-chain α or β hemiacetal form which is written using a Haworth structure. In maltose, two glucose units are linked by an acetal or glycosidic bond.

$$\begin{array}{c} CHO \\ | \\ HCOH \\ | \\ HOCH \\ | \\ HCOH \\ | \\ HCOH \\ | \\ CH_2OH \end{array}$$

D-Glucose
(Fischer open-chain form)

α-D-Glucose
(Haworth structure)

Maltose
(Haworth structure)

Reducing Sugars

All of the monosaccharides and most of the disaccharides are easily oxidized. While the monosaccharide exists as a hemiacetal most of the time, there is always a small amount that reverts to the open-chain form which contains the aldehyde group. The aldehyde group is oxidized by Benedict's reagent.

When the functional group of a sugar is oxidized, another reactant, in this case Cu^{2+}, is reduced. The name **reducing sugar** is given to a sugar that causes this reduction. In this experiment, you will use Benedict's reagent to test for the presence of reducing sugars among the mono- and disaccharides. When the cupric ion, Cu^{2+}, reacts with a reducing sugar, it is reduced to Cu^+ and forms a red precipitate of cuprous oxide, $Cu_2O,(s)$.

$$\underset{\text{reducing sugar}}{\text{sugar-}\overset{O}{\overset{\|}{C}}H} + \underset{\text{(blue)}}{2\ Cu^{2+}} \xrightarrow{\text{heat}} \underset{\text{oxidized sugar}}{\text{sugar-}\overset{O}{\overset{\|}{C}}OH} + \underset{\text{(red-orange)}}{Cu_2O(s)}$$

Sucrose is a disaccharide that does not react with Benedict's reagent. In sucrose, the monosaccharides of glucose and fructose are linked through the oxygen atoms of the hemiacetal parts of the molecules. Therefore, there is no remaining hemiacetal that can revert to the open-chain form that would cause the reduction of the cupric ion.

CARBOHYDRATES

Seliwanoff's Test

Seliwanoff's test can be used to distinguish between ketohexoses and aldohexoses. When ketohexoses are present, a deep red color is formed rapidly. Aldohexoses give a light pink color which develops more slowly.

Iodine Test for Polysaccharides

Starch, a polysaccharide, contains two polymers, amylopectin and amylose. Both consist of many glucose units with amylose being a straight-chain polymer and amylopectin a branched-chain polymer. When iodine reagent is added to amylose, a deep, blue-black color appears. Amylopectin and other polysaccharides give red to brown colors with iodine. The monosaccharides and disaccharides are not reactive with iodine.

Hydrolysis of Disaccharides and Polysaccharides

Disaccharides hydrolyze in the presence of an acid to give their individual monosaccharides. Polysaccharides such as amylose in starch can be hydrolyzed to smaller polysaccharides (dextrins), and to maltose. Complete hydrolysis produces many glucose molecules. This same type of hydrolysis of starch is also catalyzed by salivary and pancreatic enzymes during the digestion process.

$$\text{Sucrose} + H_2O \xrightarrow{H^+} \text{Glucose} + \text{Fructose}$$

$$\text{Amylose} + H_2O \xrightarrow{H^+} \text{Dextrins} + \text{Maltose} \xrightarrow{H^+} \text{Glucose}$$

Fermentation

Some sugars undergo fermentation in the presence of yeast, which contains the enzyme zymase. The products of fermentation are ethyl alcohol and carbon dioxide. The formation of bubbles of carbon dioxide is used as confirmation of the fermentation process.

$$C_6H_{12}O_6 + O_2 \xrightarrow{\text{yeast}} 2\ C_2H_5OH + 2\ CO_2(g)$$

LABORATORY ACTIVITIES

A. STRUCTURAL FORMULAS FOR CARBOHYDRATES

Monosaccharides Draw the Fischer projection and the Haworth formulas for α-D-glucose, α-D-fructose, and ß-D-galactose.

Disaccharides Draw the Haworth cyclic formulas for lactose, maltose and sucrose. Indicate the glycosidic bond.

Polysaccharides Draw a representative formula for amylose indicating the typical glycosidic bonds.

EXPERIMENT 34

B. BENEDICT'S TEST FOR REDUCING SUGARS

Be sure you have your safety goggles on.

Place 5 mL of Benedict's reagent in seven separate test tubes. Add 10 drops of each of the following solutions.

1% glucose, fructose, lactose, sucrose, maltose, starch, and water

Label and place all test tubes in a boiling water bath for 10 minutes. Record any changes in the blue color in the test tubes. The formation of a greenish to reddish-orange color indicates the presence of a reducing sugar. Record your results. Classify each as a reducing or nonreducing sugar.

C. SELIWANOFF'S TEST FOR KETOHEXOSES

Place 5 mL of Seliwanoff's reagent in six separate test tubes.

CAUTION: WEAR GOGGLES. *Seliwanoff's reagent contains HCl.*

Add 10 drops of each of the following solutions:

glucose, fructose, sucrose, maltose, starch, and water

Place the test tubes in a boiling hot water bath. After 1 min, observe the colors in the test tubes. Record your results. The rapid formation of a bright-red product indicates that a ketohexose is present. Identify the compounds that are ketohexoses.

D. IODINE TEST FOR POLYSACCHARIDES

Place 2 mL of each solution in five separate test tubes.

1% glucose, sucrose, maltose, starch, and water

Add 1 drop of iodine solution to each sample. Record your results. A dark blue-black color is a positive test for amylose in starch. A red or brown color indicates the presence of other polysaccharides. Indicate the compounds that are polysaccharides.

E. HYDROLYSIS OF DI- AND POLYSACCHARIDES

Place 5 mL of sucrose solution and 5 mL of starch solution in separate test tubes. Add 2 mL of 6M HCl to each sample. Label each test tube and place in a boiling water bath and heat for 10 min. Remove the test tubes and let them cool. Neutralize each sample with 6M NaOH (about 20 drops) until a drop of the mixture turns litmus paper blue.

CARBOHYDRATES

E.1   Transfer a small amount (about 1 mL) of each solution into a spot plate or on a watch glass. Add 1 drop of iodine reagent. Record observations. Compare your results to those obtained in part D and account for any differences in test results.

E.2   Add 5 mL of Benedict's reagent to the remaining hydrolyzed samples. Place in a boiling water bath for 10 minutes. Record your results and compare with the previous results of sucrose and starch with Benedict's reagent. Account for the differences observed.

F. FERMENTATION TEST

Fill fermentation tubes with samples of the following carbohydrate solutions. (See Figure 34.1.)

    1% glucose, fructose, sucrose, lactose, maltose, starch

Figure 34.1   *Fermentation tube.*

If fermentation tubes are not available, carbohydrate solutions may be placed in six test tubes. Place smaller test tubes upside down in the larger test tubes. Place your hand firmly over the mouth of the test tube and invert. When the small test tube inside has completely filled with the mixture, return the larger test tube to an upright position. (See Figure 34.2.)

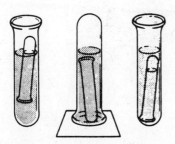

Figure 34.2   *Test tubes used as fermentation tubes.*

EXPERIMENT 34

Add 0.5 g yeast to each solution and mix well. Set the tubes aside. At the end of the laboratory period, and again at the next laboratory period, look for gas bubbles in the fermentation tubes. Record your observations.

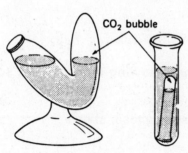

Figure 34.3 Fermentation tubes with CO_2 bubble

G. TESTING AN UNKNOWN CARBOHYDRATE

Use the carbohydrate tests to identify an unknown carbohydrate sample as a mono-, disaccharide, or polysaccharide. The possible compounds are glucose, fructose, sucrose, maltose, and amylose. Record the results for each test and your conclusions.

H. TESTING FOODS FOR CARBOHYDRATES

Obtain one of the sugar samples, a syrup sample, and two foods to test. Perform the Benedict's test and the iodine test on each. Describe the kinds of carbohydrates you can identify in each sample.

EXPERIMENT 34
CARBOHYDRATES
LABORATORY REPORT

A. STRUCTURAL FORMULAS FOR CARBOHYDRATES

Monosaccharide	Fischer	Haworth
α-D-Glucose		
α-D-Fructose		
β-D-Galactose		

EXPERIMENT 34

Disaccharide	Haworth structure
lactose	
maltose	
sucrose	

Polysaccharide	Haworth structure
amylose	

NAME_____ SECTION_____ DATE_____

B. BENEDICT'S TEST FOR REDUCING SUGARS

Carbohydrate	Observations	Reducing sugar?
glucose		
fructose		
lactose		
sucrose		
maltose		
starch		
water		

C. SELIWANOFF'S TEST FOR KETOHEXOSES

Carbohydrate	Observations	Ketohexose?
glucose		
fructose		
sucrose		
maltose		
starch		
water		

EXPERIMENT 34

D. IODINE TEST FOR POLYSACCHARIDES

Carbohydrate	Observations	Polysaccharide?
glucose		
sucrose		
maltose		
starch		

E. HYDROLYSIS OF DI - AND POLYSACCHARIDES

Carbohydrate	Iodine test (E.1)	Benedict's test (E.2)
sucrose (hydrolyzed)		
starch (hydrolyzed)		

a. Compare the results of the iodine test for starch before and after hydrolysis. Why is there a difference?

b. Compare the results of the Benedict's test with hydrolyzed sucrose with the results in part B before hydrolysis. Why are they different?

c. Compare the results of the Benedict's test with hydrolyzed starch with the results in part B before hydrolysis. Why are they different?

NAME_____ SECTION_____ DATE_____

F. FERMENTATION TEST

Carbohydrate	Observations	Gas bubbles?
glucose	_____	_____
fructose	_____	_____
sucrose	_____	_____
lactose	_____	_____
maltose	_____	_____
starch	_____	_____

What gas accumulated in the test tubes? _____

G. TESTING AN UNKNOWN CARBOHYDRATE

Unknown number _____

Test	Observations	Conclusions
Benedict's test	_____	_____
Seliwanoff's	_____	_____
Iodine	_____	_____
Fermentation	_____	_____

Identification of Unknown _____

Give reasoning.

EXPERIMENT 34

H. TESTING FOODS FOR CARBOHYDRATES

Food item	Benedict's test	Iodine test
_____	_____	_____
_____	_____	_____
_____	_____	_____
_____	_____	_____

List some types of carbohydrates found in the foods you tested.

QUESTIONS AND PROBLEMS

Indicate whether the following carbohydrates will give a positive (+) reaction or negative (-) reaction in each test listed.

Carbohydrate	Benedict's	Tests Seliwanoff's	Iodine	Fermentation
Glucose	_____	_____	_____	_____
Fructose	_____	_____	_____	_____
Galactose	_____	_____	_____	_____
Sucrose	_____	_____	_____	_____
Lactose	_____	_____	_____	_____
Maltose	_____	_____	_____	_____
Amylose	_____	_____	_____	_____

EXPERIMENT 35
SAPONIFICATION OF LIPIDS

GOALS

1. Observe some physical and chemical properties of lipids.
2. Distinguish between saturated and unsaturated fats.
3. Prepare a soap by the saponification of a triglyceride.
4. Observe the reactions of soap with soft water, oil and $CaCl_2$.

MATERIALS NEEDED

fatty acids: stearic acid, oleic acid
lipids: olive oil, safflower oil, cholesterol, lecithin,
    crisco or shortening, lecithin, vitamin A or cod liver oil
test tubes
2% Br_2 in 1,1,1-trichloroethane
methylene chloride, CH_2Cl_2
solid fat for soap: lard, Crisco or commercial shortening, tallow
125-mL Erlenmeyer flask
boiling chips
95% ethanol
20% NaOH
aluminum foil
400-mL beaker
saturated NaCl solution
Buchner filtration apparatus
filter paper
commercial detergent or soap
pH paper
5% $CaCl_2$
5% $MgCl_2$
5% $FeCl_3$

CONCEPTS TO REVIEW

lipids
fatty acids
triglycerides
saponification

EXPERIMENT 35

BACKGROUND DISCUSSION

Lipids are a family of compounds that are grouped by similarities in solubility rather than structure. As a group, lipids are more soluble in nonpolar solvents such as ether, chloroform, or benzene. Most are not soluble in water. Important types of lipids include the fats and oils, phospholipids, terpenes and the steroids. Compounds classified as lipids include fat-soluble vitamins A, D, E and K, cholesterol, hormones, portions of cell membranes and vegetable oils.

Triglycerides

The triglycerides in fats and oils are esters of glycerol and fatty acids which are carboxylic acids that are typically 14 to 18 carbons in length. When the fatty acid contains double bonds, the triglyceride is referred to as an *unsaturated fat*. When the fatty acid consists of a alkane-like carbon chain, the triglyceride is a *saturated fat*.

Fatty Acids

$$CH_3(CH_2)_{16}COH$$

Stearic acid
(saturated)

$$CH_3(CH_2)_7CH=CH(CH_2)_7COH$$

Oleic acid
(unsaturated)

$$CH_2-O-C-(CH_2)_{16}CH_3$$
$$CH-O-C-(CH_2)_{16}CH_3$$
$$CH_2-O-C-(CH_2)_{16}CH_3$$

Tristearin, a saturated fat

Typically the saturated fats are solid at room temperature. Fats containing unsaturated fatty acids are usually liquids at room temperature.

Bromine Test for Unsaturation

The presence of unsaturation in a fatty acid or a triglyceride can be detected by the bromine test performed in an earlier experiment for double bonds. If the orange color of the bromine solution fades quickly, an addition reaction has occurred and the oil or fat is unsaturated.

Saponification: Making a Soap

The hydrolysis of a fat or oil by a base such as NaOH is called *saponification*. The products are glycerol and the salts of the fatty acids which are *soaps*. The soap is precipitated out by the addition of a saturated NaCl solution. Several tests will be made with the soap you prepare such as measuring its pH, its ability to form suds in soft and hard water, and its reaction with oils.

$$\begin{array}{c}CH_2-O-\overset{O}{\overset{\|}{C}}-R\\ CH-O-\overset{O}{\overset{\|}{C}}-R\\ CH_2-O-\overset{O}{\overset{\|}{C}}-R\end{array} + 3\ NaOH \longrightarrow \begin{array}{c}CH_2OH\\ CHOH\\ CH_2OH\end{array} + 3\ R-\overset{O}{\overset{\|}{C}}-O^-\ Na^+$$

Triglyceride   Glycerol   Soap (sodium salts of fatty acids)

A soap has a dual nature. The nonpolar carbon chain which is hydrophobic is attracted to nonpolar substances such as grease and the polar carboxylate salt which is hydrophilic is attracted to water. In hard water, the carboxylate ends of soap react with Ca^{2+}, Fe^{3+} and/or Mg^{2+} ions and form an insoluble substance which we see as a gray line in the bath tub or sink.

LABORATORY ACTIVITIES

   WEAR SAFETY GOGGLES!

A. SOLUBILITY OF SOME LIPIDS

Place about 2 mL of water in each of six test tubes. Add 10 drops, or the amount of solid lipid that you obtain on the tip of a microspatula.

  stearic acid, safflower oil, Crisco (or shortening)
  lecithin, cholesterol, vitamin A (puncture capsules or use cod liver oil)

Stopper and shake each test tube. Record your observations. Repeat the solubility tests using about 2 mL of hexane.

B. TEST FOR UNSATURATION

CAUTION: AVOID CONTACT WITH THE BROMINE SOLUTION; IT CAN CAUSE PAINFUL BURNS. WEAR GOGGLES AND WORK IN THE HOOD.

Place 10 drops, or a small amount of solid, of the following in separate test tubes:

stearic acid, oleic oil, safflower oil, olive oil

To dissolve solids, add 1 mL methylene chloride, CH_2Cl_2. Add the 2% bromine solution, drop by drop, until a permanent orange color remains. Count the number of drops of bromine added. Record.

C. SAPONIFICATION: PREPARATION OF SOAP

CAUTION: OIL AND ETHANOL WILL BE HOT, AND MAY SPLATTER OR CATCH FIRE. KEEP A WATCH GLASS NEARBY TO SMOTHER FIRE. NaOH IS CAUSTIC AND CAN CAUSE PERMANENT EYE DAMAGE. WEAR GOGGLES AT ALL TIMES.

Place 5 g of a solid fat (lard, shortening, or tallow) in a 125-mL Erlenmeyer flask. Add 20 mL 95% ethyl alcohol and 20 mL of 20% NaOH. *Use care in its use*. Add a few boiling chips to the mixture to help prevent excessive foaming.

Cover the mouth of the flask with aluminum foil. Poke a hole in the foil. Place the flask in a hot water bath using a 400-mL beaker approximately half-full of boiling water. Heat for 30 minutes or until the solution becomes clear. There should be no separation of layers in the flask. If foaming is excessive, *reduce* the heat.

To test for completeness of saponification, carefully transfer a few drops of the saponified solution to 5 mL of water in a test tube. If fat droplets form, add 5 mL 20% NaOH and 5 mL ethanol to the flask and heat carefully for 10 more minutes. When saponification is complete (solution is clear), let the mixture cool.

Obtain 100 mL of a saturated NaCl solution (30 g NaCl in 100 mL water). Pour the soap mixture into the salt solution. Solid soap should float at the surface. Collect the solid soap, using a Buchner funnel and suction filtration apparatus. Wash the soap with two 15 mL portions of distilled water. Use plastic gloves to carefully pack the solid pieces of soap together to form a small cake.

HANDLE CAREFULLY: *The soap may still contain NaOH which can irritate the skin.*

D. PROPERTIES OF SOAPS

Prepare a soap solution by dissolving 1 g of your soap in 50 mL of distilled water. Also prepare a solution of a commercial soap or detergent (Tide, All, Cheer, etc.) in the same amounts. (Use 20 drops of a liquid soap.)

SAPONIFICATION OF LIPIDS

D.1 Place 5 mL of your soap solution and 5 mL of a commercial soap solution in separate test tubes. Stopper and shake for 10 seconds. Layers of foam should form. Describe the results.

D.2 Using pH paper, determine the pH of each soap solution. Record results.

D.3 Place 5 ml of each soap solution in separate test tubes. Add 10 drops of safflower oil to each. Stopper and shake each test tubes. Observe the contents of each test tube. Is the oil dispersed? Record the results.

D.4 Place 5 mL of your soap solution in four separate test tube. Place 2 mL 5% $CaCl_2$ in the first; 2 mL of 5% $MgCl_2$ in the second; 2 mL of 5% $FeCl_3$ in the third; and 2 mL of water in the fourth. Stopper and shake. Record your observations.

NAME_____ SECTION_____ DATE_____

EXPERIMENT 35
SAPONIFICATION OF LIPIDS
LABORATORY RECORD

A. SOLUBILITY OF SOME LIPIDS

Compound	Observations in water	in hexane
Stearic acid		
Safflower oil		
shortening		
Lecithin		
Cholesterol		
Vitamin A		

Explain why the above compounds are classified as lipids.

What type of solvent is needed to remove an oil spot? Why?

EXPERIMENT 35

B. TEST FOR UNSATURATION

Compound	Number of drops Br_2 solution	Saturated/unsaturated?
Fatty acids		
Stearic acid		
Oleic acid		
Triglycerides		
Safflower oil		
Olive oil		

a. Write the structural formulas of stearic acid and oleic acid.

    Stearic acid   Oleic acid

b. Which is the more unsaturated fatty acid? _____

   Which is the more saturated fatty acid? _____

c. Why is the melting point of stearic acid (70°C) higher than the melting point of oleic acid (4°C)?

d. Which triglyceride is the more unsaturated? Why?

NAME_____ SECTION_____ DATE_____

C. SAPONIFICATION: FORMATION OF SOAP

A solid fat used in the saponification will contain mostly tristearin. Write an equation for saponification of tristearin with NaOH that produced your soap.

D. REACTIONS OF SOAPS

	Observations	
	soap (prepared)	commercial product _____
D.1 Shaking	_____	_____
D.2 pH	_____	_____
D.3 Oil	_____	_____

What is the effect of a soap on a layer of oil?

D.4 Observations with hard water ions

$CaCl_2$ _____

$MgCl_2$ _____

$FeCl_3$ _____

How does soap react with Ca^{2+}, Mg^{2+}, and Fe^{3+} ions?

EXPERIMENT 36
AMINO ACIDS

GOALS

1. Identify the side (R) group in amino acids.
2. Use paper chromatography to separate amino acids in a mixture.
3. Calculate R_f values for amino acids.
4. Identify amino acids by their R_f values.
5. Use R groups to determine if an amino acid will be acidic, basic, or neutral; hydrophobic or hydrophilic.

MATERIALS NEEDED

amino acids (1% solutions): alanine, glutamic acid, serine, aspartic acid, lysine, phenylalanine, unknown amino acids

Aspartame (artificial sweetener)
150 mL beaker
400 or 600 mL beaker
6 M HCl
plastic wrap
chromatography paper, Whatman #1 (12 cm x 24 cm)
solvent: n-butanol: glacial acetic acid: water (3:1:1)
toothpicks, or capillary tubing
drying oven (100°C)
0.2% ninhydrin spray
hair dryer (optional)
metric ruler

CONCEPTS TO REVIEW

amino acids

BACKGROUND DISCUSSION

We obtain amino acids in our diet by hydrolyzing proteins. In our cells, amino acids are used to build tissues, enzymes, skin and hair. Amino acids are very similar in structure because each has an amino group and a carboxylic acid group. Individual amino acids have different organic groups called *R groups*, attached to the alpha carbon atom. Variations in the R groups determine whether an amino acid is hydrophilic or hydrophobic, acidic, basic, or neutral. Table 36.1 lists the R groups for some amino acids.

EXPERIMENT 36

```
                R  O
                |  ||
amino           |  ||        carboxylic acid group
group    NH₂—C—C—OH
                |
                H
```

Table 36.1   Some Typical Amino Acids

R	Amino Acid	Abbreviation
H—	Glycine	Gly
CH₃—	Alanine	Ala
HO—C(=O)-CH₃-CH₂—	Glutamic acid	Glu
HO—C(=O)-CH₂—	Aspartic acid	Asp
HO-CH₂—	Serine	Ser
C₆H₅-CH₂—	Phenylalanine	Phe
SH-CH₂—	Cysteine	Cys
(CH₃)₂CH—	Valine	Val
HO-C₆H₄-CH₂—	Tyrosine	Tyr
(indole)-CH₂—	Tryptophan	Trp
NH₂-CH₂-CH₂-CH₂-CH₂—	Lysine	Lys

DIPEPTIDES

A dipeptide is formed when two amino acids bond together by forming a peptide bond; the amino group of one amino acid bonds with the carboxylic acid group of the next amino acid. Many amino acids can link together to form a polypeptide or protein. The specific order of amino acids determines the type of protein that results.

Amino acid (1) +   Amino acid (2) ⟶ Dipeptide + H_2O

$$NH_2-\underset{R_1}{CH}-\underset{\|}{\overset{O}{C}}-OH + H_2N-\underset{R_2}{CH}-\underset{\|}{\overset{O}{C}}-OH \longrightarrow NH_2-\underset{R_1}{C}-\underset{\|}{\overset{O}{C}}-\underset{\text{peptide bond}}{NH}-\underset{R_2}{C}-\underset{\|}{\overset{O}{C}}-OH + H_2O$$

Chromatography of Amino Acids

Paper chromatography separates amino acids in a mixture. A drop of the mixture is placed on lower edge of the paper. Then the spotted edge of the paper is placed in a container with a layer of solvent. The paper acts as a wick and the solvent flows up the chromatogram. As the solvent travels up the paper, it carries the amino acids with it. Since the amino acids have different solubilities in the solvent and in the stationary phase (paper), they will separate from each other. Those that are attracted more to the solvent will move higher on the paper. The amino acids can be detected by spraying a dried chromatogram with ninhydrin to make the amino acids visible.

After visualization of the amino acids, the distance they traveled from the starting line can be measured. The relationship of the distance traveled by an amino acid compared to the distance traveled by the solvent is called the R_f value.

$$R_f = \frac{\text{distance traveled by amino acid}}{\text{distance traveled by solvent}}$$

When an unknown amino acid is to be identified, its R_f value and color with ninhydrin can be matched to the R_f values and colors of known amino acids. In this way, the amino acids present in an unknown mixture of amino acids or a hydrolysate of a dipeptide such as Aspartame can be identified.

LABORATORY ACTIVITIES

Wear protective goggles!

A. SIDE (R) GROUPS OF AMINO ACIDS

A.1   Write the structural formula for each of the amino acids listed.

A.2   Indicate whether each amino acid would be hydrophobic or hydrophilic.

A.3   Indicate whether the isoelectric point of each amino acid will be at a pH that is acidic, basic, or neutral.

EXPERIMENT 36

B. PAPER CHROMATOGRAPHY

Preparation of Dipeptide Hydrolysate (OPTIONAL)

Place 1 g Aspartame in a 125-mL Erlenmeyer flask. Add 30 mL of 6M HCl and place the flask in a boiling water bath. Use a clamp to support the flask. Heat at a gentle boil for 30 minutes.

Preparation of Chromatography Tank

     WORK IN THE HOOD: AVOID FUMES.

Pour about 15 mL of the butanol:HAc:water solvent into 400 or 600 mL beaker. The depth of the solvent mixture should be about 1 cm and must not exceed the origin line on your chromatogram. Cover the beaker with plastic wrap. Label the beaker with your name and leave in the hood.

Preparation of Paper Chromatogram

Obtain a piece of Whatman No. 1 chromatography paper (12 cm x 24 cm). (The paper should fit into the tank when it forms a cylinder.) Keep your fingers off the paper as much as possible since you will transfer amino acids from your skin. Handle the paper at the edge or with plastic gloves or forceps. Draw a pencil (lead) line 2 cm from the long edge of the paper. This is the starting or origin line. Mark and number eight points, about 2 cm apart, along the line. (See Figure 36.1) Place your name or initials in the upper corner with the pencil.

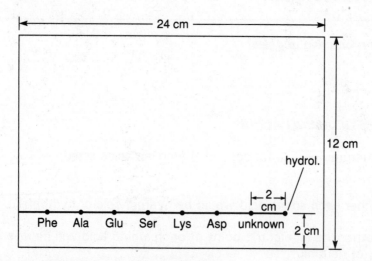

Figure 36.1 *Preparation of a chromatogram.*

Application of Amino Acids

Using the toothpick applicators or capillary tubes provided in each amino acid solution, lightly touch the tip to the paper. You may retouch the spot to apply some more, but keep the diameter of the spot as small as possible, like the size of the letter *o*. Always return the applicator to the same amino acid solution. Label each spot as you go along, using a lead pencil. Also apply a spot of the unknown amino acid and a spot of the dipeptide hydrolysate from Aspartame. A hair dryer can be used to dry the spots. When dry, repeat the application of amino acids, unknown, and protein hydrolysate two more times.

Running the Chromatogram

Form a cylinder with the paper. Join the edges with staples without overlapping the paper. Place the paper cylinder in the chromatography tank with the spotted edge down, making sure that the cylinder does not touch the sides. Make sure that the level of solvent is below the origin line you marker on the chromatogram. See Figure 36.2. Cover the beaker with the plastic wrap. Do not disturb the tank as the solvent flows up the chromatogram until the solvent line is about 1-2 cm from the top edge of the paper. *DO NOT LET THE SOLVENT RUN OVER THE TOP OF THE PAPER*. It may take from 30 to 60 minutes.

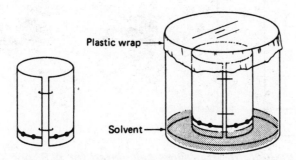

Figure 36.2 Chromatogram in solvent tank.

Visualization of Amino Acids

When the solvent line is about 1-2 cm from the top of the chromatogram, remove the paper carefully. Take out the staples and spread the chromatogram out on a paper towel IN THE HOOD WITH VENTILATION. **Mark the solvent line immediately using a pencil**. Place the chromatogram on a paper towel and let it dry. A hair dryer may be used to speed up the drying process. When the paper is dry, set it against a board in the hood, and spray lightly, but evenly with 0.2% ninhydrin spray.

 CAUTION: USE THE NINHYDRIN SPRAY INSIDE THE HOOD USING A LIGHT SPRAY. DO NOT BREATHE FUMES OR GET SPRAY ON YOUR SKIN.

EXPERIMENT 36

B.1 The sprayed paper can be dried by placing it in a drying oven for 3-5 minutes. Colored spots will appear where the ninhydrin reacted with the amino acids. Outline each spot with a pencil. Place a dot at the center of each spot. Record the color of each spot.

B.2 *Calculation of R_f Values* The R_f value represents the ratio of the distance traveled by an amino acid compared to the distance traveled by the solvent. Measure the distance(mm) from the starting line (origin) to the center dot of each spot to obtain the distance traveled by each amino acid. Measure the distance from the starting line to the solvent line to obtain the distance traveled by the solvent. See Figure 36.3.

B.3 Calculate and record the R_f values for the known amino acid samples, the unknown amino acid, and the amino acids that appear in the dipeptide hydrolysate.

$$R_f = \frac{\text{distance traveled by amino acid}}{\text{distance traveled by solvent}}$$

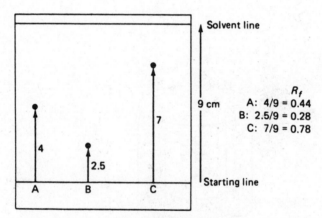

Figure 36.3 A developed chromatogram. (R_f values calculated for A, B, and C.)

B.4 *Identification of Unknown Amino Acids* Compare the color and R_f values produced by the unknown amino acid and the hydrolysate. Identical amino acids will travel the same distance, give similar R_f values and form the same color with ninhydrin. Identify the amino acids in the unknown and in the dipeptide.

NAME_____ SECTION_____ DATE_____

EXPERIMENT 36
AMINO ACIDS
LABORATORY REPORT

A. pH OF AMINO ACIDS

Amino Acid	Structure	Hydrophobic/hydrophilic	Acidic, Basic, or Neutral
Alanine			
Lysine			
Glutamic acid			
Serine			
Aspartic acid			

EXPERIMENT 36

B. PAPER CHROMATOGRAPHY

B.1 Attach or sketch the results of the paper chromatogram:

B.2 Distance from origin to final solvent line _____

B.3

Amino Acid	Color	Distance Traveled	R_f value
Alanine			
Lysine			
Glutamic acid			
Serine			
Aspartic acid			
Phenyl-alanine			
Unknown			
Aspartame hydrolysate			

B.4 Amino acid(s) in unknown _____

Amino acids in dipeptide hydrolysate _____

EXPERIMENT 37
PROTEINS

GOALS

1. Identify the structural patterns of proteins.
2. Use the isoelectric point of casein in milk to isolate the protein.
3. Use chemical tests to identify amino acids and proteins.
4. Observe the denaturation of proteins.

MATERIALS NEEDED

1% amino acid solutions: glycine, alanine, cysteine, tyrosine
1% solutions of proteins: gelatin, egg albumin

10% HNO_3
10% NaOH
250-mL beaker
nonfat milk
hot water bath
thermometer
10% acetic acid
Buchner filtration apparatus
watch glass

0.5% $CuSO_4$
pH paper
0.2% ninhydrin solution
con. HNO_3
1% $PbAc_2$
1% $AgNO_3$
ethyl alcohol

CONCEPTS TO REVIEW

amino acids
peptide bonds
structural levels of proteins
denaturation

BACKGROUND DISCUSSION

Proteins are composed of about twenty amino acids, ten of which must be obtained in the diet (essential amino acids). Proteins have specific, three-dimensional structures determined by the order of amino acids. The peptide bonds that bond the amino acids together are the first or primary level of protein structure.

EXPERIMENT 37

Amino acid (1) + Amino acid (2) ⟶ Dipeptide

$$NH_2-\underset{\underset{R_1}{|}}{CH}-\underset{\underset{O}{\|}}{C}-OH \quad + \quad H_2N-\underset{\underset{R_2}{|}}{C}-\underset{\underset{O}{\|}}{C}-OH \quad \longrightarrow \quad NH_2-\underset{\underset{R_1}{|}}{C}-\underset{\underset{O}{\|}}{C}-NH-\underset{\underset{R_2}{|}}{C}-\underset{\underset{O}{\|}}{C}-OH$$
peptide bond

Secondary structures include the alpha(a)-helix formed by the coiling of the peptide chain, or a pleated sheet structure formed between protein strands. In the tight, compact shape of globular proteins, a tertiary structure is formed by the interaction of the side groups such as salt bridges, hydrophobic bonds, and disulfide bonds. The quaternary structure consists of an active protein that has a grouping of two or more tertiary units. Proteins make up many important features in the body including skin, muscle, cartilage, hair, fingernails, enzymes, hormones, and endorphins.

Tests for amino acids and proteins

In this experiment, we will observe reactions of amino acids and proteins that give specific colors that can be used to determine the presence of amino acids or proteins.

Biuret Test

The biuret test is positive for peptides or proteins with two or more peptide bonds. A pink-violet color appears when a basic solution of Cu^{2+} is added to a peptide containing three or more amino acids. Individual amino acids will not give the biuret test, and the solution will remain blue (negative).

Ninhydrin Test

The ninhydrin test produces a blue-violet color with amino acids and most proteins. Proline and hydroxyproline give a yellow color.

Xanthoproteic Test

Amino acids or proteins containing aromatic rings with amino or hydroxyl groups such as tyrosine will react with concentrated nitric acid to give nitro-substituted benzene rings which appear as yellow-colored products.

Sulfur Test

Amino acids or proteins contain sulfur in their side chains (such as cysteine) which react with lead acetate ($PbAc_2$) to form a black precipitate of PbS(s).

Denaturation of Proteins

Denaturation of a protein occurs when the secondary or tertiary structure of the protein is altered or destroyed. In many cases, coagulation of the protein occurs. Agents causing denaturation include heat, acid, base, ethanol, tannic acid, and heavy metal ions of silver, lead, and mercury. Heat increases the motion of the atoms which disrupts the hydrogen bonding and hydrophobic (nonpolar) bonds. Strong acids add hydrogen ions to ionic side groups, and strong bases remove hydrogen ions from the ionic side groups. As a result, salt bridges (ionic bonds) are broken. Alcohol is an organic solvent that disrupts the hydrogen bonding of the secondary and tertiary protein structures. The heavy metal ions, Ag^+, Pb^{2+}, and Hg^{2+}, react with the disulfide bonds of the tertiary structures and the carboxylic groups of the acidic R groups.

LABORATORY ACTIVITIES

Wear eye protection during this experiment!

A. SEPARATION OF A PROTEIN (CASEIN) FROM MILK

A typical source of protein is milk, which contains the protein casein. When nonfat milk is acidified and the isoelectric point is reached, the protein separates out of the solution. The amount of protein will be determined and the percentage of casein in milk calculated.

A.1 Weigh a 250-mL beaker. Record the mass. Add about 50 ml of nonfat milk to the flask and reweigh. This gives the mass of the nonfat milk.

A.2 Set the beaker on a hot plate, or warm gently with a burner until the temperature reaches about 50°C. Remove the heat source. Slowly add 10% acetic acid dropwise. When the isoelectric point is reached, the casein becomes insoluble, and precipitates out (coagulates) leaving a clear solution. When no further precipitation occurs, stop adding acid. If the liquid layer does not clear, heat the mixture gently for a few more minutes.

Use the Buchner filtration apparatus to collect the solid protein. Place the protein on a paper towel to remove excess water. Weigh a watch glass to the nearest 0.1 g. Place the dried protein on the watch glass, reweigh and record. This will give the mass of protein from the milk sample.

A.3 Calculate the percentage of casein in the nonfat milk.

$$\% \text{ casein} = \frac{\text{mass (g) of casein}}{\text{mass (g) of milk}} \times 100\%$$

B. BIURET TEST FOR PROTEINS

Place 2 mL of a 1% solution, or a small amount of solid on the tip of a spatula, in five separate test tubes:

> glycine, alanine, gelatin, egg albumin, and casein (from part A)

Add 2 mL of 10% NaOH and mix thoroughly. To each, add 2 drops of 0.5% cupric sulfate ($CuSO_4$), and stir. Record the appearance of each solution. The formation of a pink-violet color indicates the presence of a protein with two or more peptide bonds. If such a protein is not present, the blue color of the cupric sulfate will remain (negative). Record the results and your conclusions.

C. NINHYDRIN TEST

Place 2 mL or a small amount of solid of the following in five separate test tubes and label:

> glycine, alanine, gelatin, egg albumin, and casein (from Part A)

Add 1 mL of 0.2% ninhydrin solution to each sample. Place the test tubes in a boiling water bath for 5 minutes. Look for the formation of a blue-violet color. Record your observations.

D. XANTHOPROTEIC TEST (optional)

Place 2 mL of the following solutions or a small amount of solid on the tip of a spatula in five separate test tubes and label:

> alanine, tyrosine, gelatin, egg albumin, and casein (from Part A)

Add 10 drops of concentrated HNO_3 to each sample. Place the test tubes in a boiling water bath and heat for 2 minutes.

CAUTION: CONCENTRATED HNO_3 and 10% NaOH ARE EXTREMELY CORROSIVE AND DAMAGING TO SKIN AND EYES. WEAR GOGGLES AT ALL TIMES. USE CARE IN HANDLING CON. HNO_3 AND NaOH.

Cool the samples by placing the test tubes in cold water and let them cool. *CAREFULLY* add 10% NaOH, drop by drop, until the solution is just basic (turns red litmus blue). Look for formation of a yellow-orange color. Record the results and your conclusions.

E. SULFUR TEST

Place 2 mL of a solution, or a small amount of solid, of the following in separate tests tubes and label:

alanine, cysteine, gelatin, albumin, and casein (from Part A)

Carefully add 4 mL of 10% NaOH to each sample. Add 3 drops of 1% $PbAc_2$ solution. Place the test tubes in a boiling-water bath for 5 minutes. Look for the formation of a black precipitate. Record the results and your conclusions.

F. DENATURATION OF PROTEINS

Prepare five test tubes each with 3 mL egg albumin solution. Use one sample for each of the following tests. Record your observations and give a brief explanation for the results.

F.1 *Effect of heat* Heat the egg albumin solution until it boils.

F.2 *Effect of strong acid* Add 2 mL of 1 M HNO_3.

F.3 *Effect of strong base* Add 2 mL of 1 M NaOH.

F.4 *Effect of alcohol* Add 5 mL 95% ethyl alcohol. Mix.

F.5 *Effect of heavy metals* Slowly add 10 drops of 1% $AgNO_3$.

NAME_____ SECTION_____ DATE_____

_____

EXPERIMENT 37
PROTEINS
LABORATORY REPORT

_____

A. SEPARATION OF A PROTEIN (CASEIN) FROM MILK

A.1   Mass of beaker + milk   _____ g

       Mass of beaker   _____ g

       Mass of nonfat milk   _____ g

A.2   Mass of watch glass + casein   _____ g

       Mass of watch glass   _____ g

       Mass of casein   _____ g

A.3   % casein in milk   _____ %
       Show calculations:

B. BIURET TEST FOR PROTEINS

Compound	Observations	Conclusions
Glycine	_____	_____
Alanine	_____	_____
Gelatin	_____	_____
Albumin	_____	_____
Casein	_____	_____

C. NINHYDRIN TEST

Compound	Observations	Conclusions
Glycine		
Alanine		
Gelatin		
Albumin		
Casein		

D. XANTHOPROTEIC TEST

Compound	Observations	Conclusions
Alanine		
Tyrosine		
Gelatin		
Albumin		
Casein		

E. SULFUR TEST

Compound	Observations	Conclusions
Alanine		
Cysteine		
Gelatin		
Albumin		
Casein		

NAME_____ SECTION_____ DATE_____

F. DENATURATION OF PROTEINS

	Agent	Observations
F.1	Effect of heat	_____
F.2	Effect of acid	_____
F.3	Effect of base	_____
F.4	Effect of alcohol	_____
F.5	Effect of heavy metal	_____

EXPERIMENT 37

QUESTIONS AND PROBLEMS

1. Write the primary structure for the polypeptide of Ala-Gly-Cys-Phe.

2. Will the polypeptide in question 1 give a positive or negative result in the following tests for a protein or amino acid? Explain why or why not.

   a. Biuret

   b. Xanthoproteic

   c. Sulfur

   d. Ninhydrin

3. When does a protein undergo denaturation?

4. Under what circumstances would the denaturation of a protein be useful?

EXPERIMENT 38
ENZYME ACTIVITY

GOALS

1. Prepare a solution of an active enzyme.
2. Determine the effects of enzyme concentration, substrate concentration, temperature, and pH upon the activity of an enzyme.
3. Observe that denaturation destroys enzymatic activity.

MATERIALS NEEDED

test tubes
test tube rack
thermometer
beakers
droppers
spot plate, or wax paper
temperature baths:
boiling water bath(100°C)
ice-water bath(0°C)
body-temperature bath (37°C)

1% starch (buffered pH 7.0)
iodine reagent
0.1 M $AgNO_3$
crushed ice
0.1 M NaCl
ethanol
buffers, pH 4, 7, 10
timer

CONCEPTS TO REVIEW

proteins
enzymes
lock-and-key theory
factors affecting enzyme activity
inhibition

BACKGROUND DISCUSSION

Enzymes are biological catalysts. They participate in every biological reaction necessary for the maintenance of a living system. An enzyme acts upon the reactants or *substrates* of a reaction to give an *enzyme-substrate* complex. A reaction occurs at the *active site* that produces *products* and the enzyme which can be used again.

Enzyme + Substrate ⇌ Enzyme-substrate complex ⟶ Enzyme + Product

E + S ⇌ ES ⟶ E + P

EXPERIMENT 38

Enzymes catalyze cellular reactions at body temperature and mild conditions. The rates of enzyme-catalyzed reactions are much faster than they would be without the enzymes. Enzymes speed up a reaction by lowering the energy required to make the reaction occur.

In this experiment, you will use a readily available enzyme, salivary α-amylase, which begins the hydrolysis of carbohydrates (amylose) in the mouth. Your own saliva containing amylase will be used for this experiment although the levels of amylase vary from one person to another. To follow the reactions of amylase with starch, you will test for the hydrolysis of starch by reacting samples of the reaction mixtures with iodine. A starch solution gives a blue-black color with iodine. The amylase catalyzes the hydrolysis of the α-1,4-glycosidic bonds to give smaller polysaccharides, dextrins, maltose, and glucose.

$$\text{Starch} \xrightarrow{\alpha\text{-Amylase}} \text{Smaller polysaccharides, dextrins, maltose} \xrightarrow{\alpha\text{-Amylase}} \text{Glucose}$$

When the starch is hydrolyzed, the blue-black color produced with iodine no longer occurs, and only the red or gold color of the iodine solution is seen. The faster the α-amylase hydrolyzes the starch, the more quickly the blue-black color fades. If the blue-black color does not fade, you may conclude that the enzyme is no longer active and that no hydrolysis of starch has occurred.

Factors Affecting Enzyme Activity

Each experiment must be timed. As you proceed with each experiment, you will check enzyme activity by reacting a few drops of the reaction mixture with iodine. The time at which the blue-black color of starch disappears will be noted in each experiment.

When enzyme activity is high, the time required for the starch to disappear will be very short. When the enzyme is operating poorly or not at all, the activity will be low and more time will be required for the starch to disappear. In some cases, the enzyme will be completely inactivated and the blue-black color will persist throughout the entire experiment. You will then assess the effects of substrate concentration, pH, temperature and heavy metals upon the relative enzyme activity.

LABORATORY ACTIVITIES

Starch-iodine Test

Prepare a spot plate or sheet of wax paper for testing starch in the samples. Place 1 drop of iodine reagent in two of the depressions of the spot plate or on the wax paper.

Place a few drops of water in the first depression with iodine. Add a few drops of the starch solution to the second depression with iodine. Record the colors of the iodine reagent with water and with starch. The reaction with starch should give a deep blue-black color.

ENZYME ACTIVITY

A. EFFECT OF ENZYME CONCENTRATION

You and your partners will need to collect about 3 mL of saliva in a small, clean beaker. Chewing gum or rubber bands may help your secretion of saliva. Add about 75 mL of distilled water to the saliva and mix.

Fill a large beaker about half full of water and warm to a temperature of about 37°C. Pour 20 mL of 1% starch in a test tube and place it in the warm water bath. Using four test tubes that will hold 15 mL of solution, place the following portions of saliva solution and water and label. Place the test tubes in the 37°C water bath for 5 minutes.

Test tube	mL of saliva solution	mL of water
1	0 (control)	10
2	2	8
3	5	5
4	10	0

Remove the test tube containing 1% starch and quickly add 5 mL to each of the four test tubes in the water bath. Stir thoroughly, and place the test tubes back into the water bath. Record the time that you place the test tubes in the 37°C water bath. Keep the temperature of the water bath close to 37°C, adding more warm water as needed.

One minute after addition of the enzyme, place a drop of each reaction mixture (use clean droppers each time), in the spot plate containing a drop of iodine solution, or on the wax paper. A blue-black color in the control indicates that starch is still present. If the color with iodine is light blue, red or gold, the starch has undergone hydrolysis. Record your observations for each of the reaction mixtures. Clean the spot plate or wax paper.

At 5 minutes after addition of enzyme, test the reaction mixtures again for starch using fresh drops of iodine solution in the spot plate or on the wax paper. Repeat again at 10, 15, and 20 minutes. Record the colors of the starch test for each reaction mixture.

B. EFFECT OF TEMPERATURE

Working with your partners, fill three 250 mL beakers about half full of tap water. Add ice to one to lower the temperature, warm one to about 37°C, and heat one to boiling.

0°C	ice-water mixture
37°C	warmed tap water (body-temperature)
100°C	boiling hot water bath

Place 5 mL of 1% starch solution in each of three test tubes. Place one test tube in each of the water baths. In three other test tubes, place 5 mL of saliva solution and place one in each water bath. After five minutes, mix the contents of the test tubes in each water bath.
Record the time.

When one minute has passed after the addition of enzyme, test each sample for starch.

EXPERIMENT 38

Place a few drops of each reaction mixture in a spot plate with a drop of iodine solution. Record the colors of the iodine test for each sample. Continue testing until the starch has been hydrolyzed in at least one reaction mixture, or test at 5, 10, 15, and 20 minutes.

C. EFFECT OF pH

Place 5 mL of buffer solutions of pH 4, 7, and 10 in three separate test tubes. Add 5 mL of saliva solution to each test tube. Label. Place the test tubes in a large beaker that is about half filled with water at 37°C. Pour 15 mL of 1% starch solution in a large test tube and place in the water bath. After five minutes, pour 5 mL of the starch mixtures into each of the buffer-saliva mixtures.

One minute after the starch and saliva are mixed, test each sample for starch using the iodine test. Repeat the test for starch every 5 minutes until the starch in at least one of the reaction mixtures is completely hydrolyzed or for 20 minutes. At some pH values, the enzyme will be inactive, and the test will be positive for starch the entire time.

D. INHIBITION OF ENZYME ACTIVITY

Prepare three test tubes with the following:

Test tube	
1	2 mL 0.1 M $AgNO_3$
2	2 ml 0.1 M NaCl
3	2 mL ethanol

Add 5 mL of saliva solution to each and label. Place all the test tubes in a 37°C water bath and leave for 5 minutes. To start the reaction, add 5 mL of starch solution to each and mix thoroughly.

One minute after starting the reaction, test each reaction mixture for starch. Place a few drops of the mixture in a spot plate with one drop of iodine solution. Record the colors. Continue to test for starch hydrolysis every five minutes for 20 minutes.

NAME _____ SECTION _____ DATE _____

EXPERIMENT 38
ENZYME ACTIVITY
LABORATORY REPORT

Starch-iodine Test

Color of iodine solution with water _____

Color of iodine solution with starch _____

A. ENZYME CONCENTRATION

Color of Starch-Iodine Test

Time after addition of enzyme	Amount of Saliva solution			
	0 mL	2 mL	5 mL	10 mL
1 minute	_____	_____	_____	_____
5 minutes	_____	_____	_____	_____
10 minutes	_____	_____	_____	_____
15 minutes	_____	_____	_____	_____
20 minutes	_____	_____	_____	_____

1. What is the substrate for salivary amylase?

2. What kind of reaction does the amylase catalyze?

3. In which reaction mixture was the enzyme most active?

4. How does enzyme activity change as the concentration of enzyme is increased?

EXPERIMENT 38

B. EFFECT OF TEMPERATURE

Color of Starch-Iodine Test

Time after addition of enzyme	Temperature °C		
	0	37	100
1 minute	_____	_____	_____
5 minutes	_____	_____	_____
10 minutes	_____	_____	_____
15 minutes	_____	_____	_____
20 minutes	_____	_____	_____

1. According to your experiment, what is the optimum temperature for salivary amylase?

2. How is amylase affected by a low temperature? Explain.

3. How was amylase activity affected by a high temperature? Explain.

NAME_____ SECTION_____ DATE_____

C. EFFECT OF pH

Color of Starch-Iodine Test

Time after addition of enzyme	pH 4	pH 7	pH 10
1 minute	_____	_____	_____
5 minutes	_____	_____	_____
10 minutes	_____	_____	_____
15 minutes	_____	_____	_____
20 minutes	_____	_____	_____

1. What was the optimum pH for the hydrolysis reaction?

2. Why does a lower or higher pH cause a change in enzyme activity?

3. In the stomach, the pH is 2 during digestion of food. What would be the optimum pH of pepsin, an enzyme that hydrolyzes protein in the stomach?

   What happens to the enzymatic activity of pepsin when it enters the small intestine where the pH is about 8?

EXPERIMENT 38

D. INHIBITION OF ENZYME ACTIVITY

Color of Starch-Iodine Test

Time after addition of enzyme	AgNO$_3$	Inhibitors NaCl	Ethanol
1 minute	_____	_____	_____
5 minutes	_____	_____	_____
10 minutes	_____	_____	_____
15 minutes	_____	_____	_____
20 minutes	_____	_____	_____

1. In which reaction mixture(s) did hydrolysis occur?

2. Which of the compounds added to the reaction mixture were inhibitors?

3. How might these compounds act as inhibitors of enzyme activity?

4. What is the purpose of using an alcohol swab prior to give an injection to a patient?

EXPERIMENT 39
VITAMINS

GOALS
1. Identify a vitamin as water or fat-soluble.
2. Compare the vitamin C content in a variety of citrus juices.
3. Determine the effect of heat upon vitamin C.

MATERIALS NEEDED

Vitamins A,B,C,D,E, folic acid, and others
test tubes
methylene chloride (CH_2Cl_2)
vitamin C tablets
Erlenmeyer flasks (250 mL)
50 mL buret
1% starch indicator
fruit juices: orange, grapefruit, powdered, etc.
iodine reagent (16 g KI + 1.5 g I_2 in 1 L of solution)

CONCEPTS TO REVIEW

vitamins
fat- and water soluble vitamins

BACKGROUND DISCUSSION

Vitamins are organic compounds required as cofactors for certain enzymes. However, vitamins are not synthesized by the body and must be provided in our diet. The deficiency of a vitamin in a diet can affect the activity of an enzyme bringing about a deficiency disease. A lack of vitamin C can cause scurvy, a condition characterized by bleeding gums and anemia, while a deficiency in vitamin A is associated with night blindness. A diet low in vitamin D can cause rickets in children, and a deficiency in B_{12} can lead to anemia. Because vitamins B and C have polar groups, they are soluble in water. Any excess in these vitamins is usually excreted by the body. Vitamins A, D, E, and K are nonpolar and are fat-soluble vitamins.

Testing for Vitamin C

Vitamin C, also called ascorbic acid, is used by the body to fight infection and repair damaged tissues. It is present in a many fresh fruits and vegetables including oranges, grapefruit, broccoli, lettuce and green peppers.

EXPERIMENT 39

Since vitamin C is a reducing agent, we can measure its presence by its reaction with iodine, I_2.

Vitamin C (ascorbic acid) + I_2 ⟶ Vitamin C oxidized + 2 HI

The indicator for this reaction is a starch solution. When iodine is present, the starch turns a deep-blue color. However, if vitamin C is present and it has reduced the iodine (I_2) to iodide (I^-), the starch does not react and no deep-blue color forms.

$$I_2 \xrightarrow{\text{Vitamin C}} 2\,I^-$$

blue-black with starch   no color with starch

In this experiment, we will first determine the number of mg of vitamin C that react with 1 mL of iodine reagent. Using the iodine reagent in the titration of citrus juice, we can determine the milligrams of vitamin C present. One sample of juice will be heated to determine the amount of vitamin C destroyed by heat.

LABORATORY ACTIVITIES

Safety goggles must be worn in the laboratory.

A. SOLUBILITY OF VITAMINS

Place a small amount of each vitamin in separate test tubes. Add 2 mL of water to each sample. Mix and observe each solution. If two layers form, the vitamin is not water soluble. If the vitamin dissolves to give a clear solution, it is water soluble. Record your observations. Repeat the procedure adding 2 mL methylene chloride (CH_2Cl_2) in place of water to each vitamin sample. Record your observations.

B. STANDARDIZATION OF VITAMIN C

B.1 Crush a vitamin C tablet and transfer to an 250 mL Erlenmeyer flask. Record the mass (mg) of the vitamin C in the tablet as stated on the label. Add 50 mL distilled water and 2 mL HAc (acetic acid) and mix. Add 10 drops of starch indicator.

B.2 Set up and fill a buret with prepared iodine solution. Record the initial level reading of the iodine solution in the buret.

CAUTION: KEEP IODINE REAGENT AWAY FROM CLOTHES AND SKIN.

Add the iodine from the buret to the vitamin C solution until you obtain a solution with a deep-blue color. This is the endpoint. Record the final reading of the iodine solution in the buret. Calculate the volume of iodine solution used in the titration.

B.3 Calculate the mass (mg) vitamin C that reacts with 1 mL iodine solution.

$$\frac{\text{mg vitamin C in tablet}}{\text{mL iodine solution}} = \text{mg vitamin C/1 mL iodine solution}$$

C. DETERMINATION OF VITAMIN C IN FRUIT JUICES

C.1 Record the type of fruit juice. Place a 25 mL sample of a fruit juice in a clean Erlenmeyer flask. If there is a lot of pulp in the juice, filter first with filter paper or cheese cloth. Add 25 mL of distilled water, 2 mL HAc (acetic acid), and 10 drops of 1% starch indicator.

If you use a powdered fruit drink such as Tang, weigh 1.0 g of the powder and place in a 250 mL Erlenmeyer flask. Add 50 mL of water to the powder, 2 mL HAc (acetic acid), and 10 drops of 1% starch indicator, and mix.

C.2 Record the initial level of iodine in the buret. Add iodine solution to the fruit juice and starch mixture until a deep-blue end-point is obtained. (Deeply colored juices may obscure this color). Record the final level of iodine solution added to reach the endpoint of the reaction. Calculate the volume of iodine used in the titration.

C.3 Calculate the mg vitamin C in the fruit juice sample. Use the value of mg vitamin C/ mL iodine solution obtained in step B.3.

$$\text{mL iodine solution (C.2)} \times \frac{\text{mg vitamin C (B.3)}}{\text{1 mL iodine solution}} = \text{mg vitamin C in sample}$$

Repeat the titration with other fruit juices available in the lab. Record the results.

EXPERIMENT 39

D. HEAT DESTRUCTION OF VITAMIN C

D.1 Place 25 mL of a juice in part C that had a high vitamin C content in each of two 250 mL Erlenmeyer flasks. Add 50 mL of water to each. Boil one sample for 10 minutes, and the other for 30 minutes. After 10 minutes, remove the first flask and place it in an ice-water bath. Add 10 drops of starch indicator. Fill a buret with iodine solution and record the initial level. Titrate with iodine solution to the deep-blue endpoint. Record the final level of iodine solution. Repeat at 30 minutes with the other sample. Calculate the volume of iodine solution used in each titration.

D.2 Calculate the mg of Vitamin C present in the heated samples.

D.3 Calculate the mg Vitamin C destroyed by heat (C.3—D.2).

NAME_____ SECTION_____ DATE_____

EXPERIMENT 39
VITAMINS
LABORATORY REPORT

A. SOLUBILITY OF VITAMINS

Vitamin	Solubility in		Water or fat-soluble
	water	CH_2Cl_2	
_____	_____	_____	_____
_____	_____	_____	_____
_____	_____	_____	_____
_____	_____	_____	_____
_____	_____	_____	_____
_____	_____	_____	_____

1. Which vitamins are water-soluble? Which ones are fat-soluble?

2. Give a reason for the difference in solubilities of vitamins.

3. Which vitamins would be excreted daily?

EXPERIMENT 39

B. STANDARDIZATION OF VITAMIN C

B.1 Mass of vitamin C in tablet _____ mg

B.2 Final buret reading _____ mL

Initial buret reading _____ mL

Volume of iodine solution _____ mL

B.3 mg of Vitamin C / 1 mL iodine solution _____ mg/mL I_2

Show calculations:

C. DETERMINATION OF VITAMIN C IN FRUIT JUICE

C.1 Sample 1: _____

C.2 Initial buret level _____ mL

Final buret level _____ mL

Volume iodine used _____ mL

C.3 Vitamin C in sample _____ mg
Show calculations:

350

NAME_____ SECTION_____ DATE_____

C.1 Sample 2: _____

C.2 Initial buret level   _____mL

     Final buret level   _____mL

     Volume iodine used   _____mL

C.3 Vitamin C in sample   _____mg
     Show calculations:

C.1 Sample 3: _____

C.2 Initial buret level   _____mL

     Final buret level   _____mL

     Volume iodine used   _____mL

C.3 Vitamin C in sample   _____mg
     Show calculations:

Rank the juices you tested in order of increasing vitamin C content.

_____   _____   _____(most)

If the daily requirement is 75 mg of vitamin C, how many milliliters (or grams) of each sample do you need to meet the minimum daily requirement?

Sample 1 _____

Sample 2 _____

Sample 3 _____

EXPERIMENT 39

D. HEAT DESTRUCTION OF VITAMIN C

Sample _____

		Time that sample was heated	
		10 minutes	30 minutes
D.1	Initial buret level	_____ mL	_____ mL
	Final buret level	_____ mL	_____ mL
	Volume of iodine solution	_____ mL	_____ mL
D.2	Vitamin C	_____ mg	_____ mg
D.3	mg vitamin C destroyed	_____ mg	_____ mg

How does heating affect the vitamin C content of a fruit juice?

If you wish to keep most of the vitamin C content of your vegetables, how would you prepare them?

NAME_____ SECTION_____ DATE_____

EXPERIMENT 40
DIGESTION OF FOODSTUFFS

GOALS

1. Identify the hydrolysis reactions in digestion.
2. Test for the hydrolysis products of carbohydrates, fats, and proteins.

MATERIALS NEEDED

test tubes
1% starch solution
iodine solution
spot plate or wax paper
Benedict's reagent
safflower oil
bile salts solution
whole milk

50-mL buret
0.1 M NaOH
2% pancreatin
pH meter or pH paper
phenolphthalein
hard boiled egg white
2% pepsin
0.1 M HCl

CONCEPTS TO REVIEW

carbohydrate
fat
protein
hydrolysis
digestion

BACKGROUND DISCUSSION

The digestive processes utilize enzymes to carry out the hydrolysis of large molecules in our food to molecules small enough to dialyze through the intestinal wall into the blood or lymph.

  Starch, a major carbohydrate in our foods, provides about 50% of our caloric intake. In order to use starch, it must be hydrolyzed into glucose molecules. Digestion of starch begins in the mouth by the action of an enzyme, salivary amylase. Hydrolysis continues in the small intestine through the action of pancreatic amylase, maltase, sucrase and lactase.

$$\text{starch (amylose)} \xrightarrow{\text{amylase}} \text{maltose} \xrightarrow{\text{maltase}} \text{glucose}$$

353

EXPERIMENT 40

Approximately 25-30% of our diet consists of lipids, primarily *fats* (triglycerides). Chemically, a fat is an ester of glycerol and fatty acids. Digestion of fats begins in the intestine with bile salts and the enzymatic action of lipases obtained from the gall bladder. The bile salts cause the fat to break up into smaller droplets (emulsification) increasing the surface area and the lipases hydrolyze the ester bonds of the fats.

$$\text{fats} \xrightarrow{\text{pancreatic lipase}} \text{glycerol + fatty acids}$$

Proteins which make up about 20-25% of our diet begin digestion in the stomach where HCl activates the proteases such as pepsin that begin the hydrolysis of peptide bonds. Other enzymes continue to hydrolyze polypeptides and dipeptide to give the hydrolysis products of protein which are amino acids.

$$\text{proteins} \xrightarrow{\text{pepsin, chymotrypsin}} \text{peptides, dipeptides} \xrightarrow{\text{dipeptidases}} \text{amino acids}$$

LABORATORY ACTIVITIES

Are your safety glasses on?

A. DIGESTION OF CARBOHYDRATES

A.1 Hydrolysis of Starch

Obtain two test tubes. In the first, collect 0.5 mL of saliva (salivary amylase). Place 0.5 mL of water in the second test tube. Add 5 mL of a 1% starch solution to each. Mix thoroughly. Prepare a spot plate or wax paper with 1 drop of iodine reagent to test each mixture for starch.

Every two or three minutes, place a few drops of each mixture in the spot plate or on a drop of iodine on the wax paper. A blue-black color indicates that starch remains in the mixture. Continue testing the mixtures until the deep-blue color for starch no longer forms in at least one of the test tubes. Record your observations. Save the mixtures for part A.2.

A.2 Test for Glucose

To test for the presence of glucose as a final product of starch digestion, add 5 mL of Benedict's reagent to each of the two test tubes and contents. Place the test tubes in a boiling water bath for 5 minutes. Record the colors that form. Benedict's is positive for glucose if the orange color of Cu_2O forms. Record your observations.

$$\text{Glucose} + 2\,Cu^+ \longrightarrow \text{Gluconic acid} + Cu_2O(s)$$
$$\qquad\qquad\quad \text{blue} \qquad\qquad\qquad\qquad\qquad\qquad \text{red-orange}$$

B. DIGESTION OF FATS

Action of Bile Salts

B.1 Place 2 mL of safflower oil in each of two test tubes. Add 8 mL water to one test tube. To the other sample, add 5 mL water and 3 mL of bile salts. Mix thoroughly. Observe the mixtures in the test tubes immediate. After 20 minutes, observe the mixtures in the test tubes again. Look for the separation of two layers, or the emulsification by the bile salts. Record your observations.

Hydrolysis by Lipase

Place 50 mL of whole milk in a 125 or 250 mL Erlenmeyer flask. Add 10 mL 2% pancreatin to the flask and mix thoroughly. Place the flask in a 37°C water bath, a 400 or 600 mL beaker about half full of water.

B.2 Carefully pour a few mL of the reaction mixture into a shell vial and measure the pH of the reaction mixture with a pH meter. (Or touch a drop of the milk mixture to pH paper and determine the pH.) Record the pH.

B.3 Set up a buret containing 0.1 M NaOH. Place 15 mL of the reaction mixture in a 125 mL Erlenmeyer flask. Add 3-5 drops of phenolphthalein. Add the 0.1 M NaOH from the buret until a permanent light pink color is obtained. This marks the endpoint. Record the number of milliliters of NaOH required to reach the endpoint.

At 30 minutes, remove another 15 mL sample of the reaction mixture. Determine the pH of the sample, and then titrate with 0.1 M NaOH. Record the number of milliliters of NaOH needed for titration. Repeat the pH measurement and titration with 0.1 M NaOH at 60 minutes.

C. PROTEIN DIGESTION

Prepare three test tubes as follows:

test tube	Solutions
1	5 mL water + 1 mL 0.1 M HCl
2	5 mL of 2% pepsin + 1 mL 0.1 M HCl
3	5 mL of 2% pepsin + 1 mL water

Add a small piece of egg white in each test tubes and place the test tubes in a 37° water bath for 1 hr. Record any changes in the egg white in each test tube.

NAME_____ SECTION_____ DATE_____

EXPERIMENT 40
DIGESTION OF FOODSTUFFS
LABORATORY REPORT

A. DIGESTION OF CARBOHYDRATES

A.1 *Hydrolysis of Starch*

Color with iodine

Time	Starch + amylase	Starch only
2 minutes	_____	_____
4 minutes	_____	_____
6 minutes	_____	_____
8 minutes	_____	_____
10 minutes	_____	_____

A.2 *Test for Glucose*

	Starch + amylase	Starch only
Color with Benedict's	_____	_____
Glucose (yes/no)	_____	_____

1. Where does starch digestion occur?

2. What are the products of carbohydrate digestion?

EXPERIMENT 40

B. DIGESTION OF FATS

Action of Bile Salts

B.1

	Oil + bile salts	Oil only
Initial observations	_____	_____
20 minutes	_____	_____

Hydrolysis by Lipase

Time	pH	Volume of 0.1 M NaOH
0 minutes	_____	_____ mL
30 minutes	_____	_____ mL
60 minutes	_____	_____ mL

1. What is the action of bile salts on an oil and water mixture?

2. What is the function of bile salts in the digestion of fats?

3. What products of fat hydrolysis would produce a change in pH?

NAME_____ SECTION_____ DATE_____

C. PROTEIN DIGESTION

	Water + HCl	Pepsin + HCl	Pepsin + H_2O
Initial appearance of egg white	_____	_____	_____
Appearance after 1 hour	_____	_____	_____
Has any digestion taken place?	_____	_____	_____

1. How does the digestion of protein take place in the stomach?

2. Why does a person with a low production of stomach HCl have difficulty with protein digestion?

3. What are the products of protein hydrolysis?

EXPERIMENT 41
CHEMICAL COMPOUNDS IN FOODS AND DRUGS

GOALS

1. Describe the chemical compounds present in food and drugs by reading the labels on containers.
2. Use reference materials to write the structural formula of the compounds.
2. Give the function and biological effects of those chemical compounds.

MATERIALS NEEDED

Processed foods, cosmetics and over-the-counter drugs
Merck Index
Handbook of Food Additives

BACKGROUND DISCUSSION

Many labels on foods and drugs contain the chemical names of the ingredients used in these products. The properties and functions of some of these compounds can be found in the Merck Index or the Handbook of Food Additives.

LABORATORY ACTIVITIES

1. Read the labels on four to five different kinds of packaged food and over-the-counter drugs. Record the names of the compounds listed on the label.

2. Using a reference, look up and write the structural formulas for at least ten compounds.

3. Record the function of each compound; list any precautions or dangers.

NAME_____ SECTION_____ DATE_____

EXPERIMENT 41
CHEMICAL COMPOUNDS IN FOODS AND DRUGS
LABORATORY REPORT

Item 1: _____

List of ingredients:

Structural formula	Function	Precautions/Dangers

EXPERIMENT 41

Item 2: _____

List of ingredients:

Structural formula	Function	Precautions/Dangers

Item 3: _____

List of ingredients:

Structural formula	Function	Precautions/Dangers

NAME_____ SECTION_____ DATE_____

Item 4: _____

List of ingredients:

Structural formula	Function	Precautions/Dangers

Item 5: _____

List of ingredients:

Structural formula	Function	Precautions/Dangers

EXPERIMENT 42
CHEMISTRY OF URINE

GOALS

1. Test urine-like specimens for pH, specific gravity, and the presence of electrolytes and organic compounds.
2. Test urine-like specimens for the presence of abnormally occurring compounds of proteins, glucose, and ketone bodies.

MATERIALS NEEDED

urine-like specimens (normal and pathological)
urinometer
pH paper
1% urease
conc. HCl
litmus paper
test tubes
beakers
0.1 M $AgNO_3$
0.1 M $BaCl_2$
ammonium molybdate solution

6 M HNO_3
flame-test wire
1 M HAc
$(NH_4)_2SO_4$(s)
5% nitroprusside reagent
con. NH_4OH
Benedict's reagent
6 M HCl

Optional: Reagent strips or tablets such as Clinistix, Clinitest, Ketostix, Albustix

BACKGROUND DISCUSSION

Examining a urine specimen can give much diagnostic information about the processes occurring within the body. The pH, the amounts of electrolytes, uric acid, and glucose can all lead to conclusions about the functioning of the kidneys, liver and the general state of health of the individual.

EXPERIMENT 42

Ions in urine

Urine is typically about 96% water. The other 4% consists of waste products being eliminated from the cells of the body to maintain proper osmotic pressure, electrolyte levels and pH. Urine normally contains the inorganic ions Cl^-, SO_4^{2-}, PO_4^{3-}, K^+, Na^+, NH_4^+, and Ca^{2+}. Organic components normally found include urea and uric acid. Urea is an end product of protein metabolism and uric acid is an end product of purine metabolism.

$$NH_2-\underset{\underset{O}{\|}}{C}-NH_2 \quad \text{urea}$$

The presence of Na^+ ion can be determined by a flame test. The present of other electrolytes will also be determined. Urea will be detected by its decomposition to ammonia (NH_3) and carbon dioxide (CO_2).

Glucose

Glucose does not normally appear in the urine. However, when there is a high level of glucose in the blood and the renal threshold is exceeded, glucose may show up in urine (glucosuria). Conditions such as diabetes mellitus and liver damage may be indicated. If glucose is present, it can be detected by Benedict's test. Test strips such as Clinitest have substances that also change color to indicate glucose.

pH and the specific gravity of urine

Urine usually has a pH around 6.0, although this varies considerably with diet and activity and can have a range from 4.6 to 8.0 at different times. Urine normally has a light yellow color derived from pigments formed by the breakdown of bilirubin formed during the destruction of red blood cells. The normal range for specific gravity is 1.005 to 1.030. In this experiment, you will test the pH of a urine sample, measure its specific gravity and note its color.

Ketone bodies

Ketone bodies such as acetone and acetoacetic acid are not normally found in the urine. However, they can appear when large amounts of fat are metabolized if there is an insufficiency of glucose in the diet or an inability to utilize glucose as in diabetes mellitus. Ketone bodies are associated with certain diets (low carbohydrate), starvation, diabetes mellitus, and liver damage.

Protein

High levels of protein (proteinuria) in the urine may indicate disease or damage to the kidneys or urinary tract. Protein can be detected by heating a portion of the urine specimen to coagulate the protein.

LABORATORY ACTIVITIES

Be sure you are wearing safety glasses!

A. COLOR, pH AND SPECIFIC GRAVITY

A.1 Obtain 50 mL of a urine-like specimen and 50 mL of a synthetic pathological urine sample. Describe the color of each specimen.

A.2 Use pH paper to determine the pH of the urine.

A.3 Determine the specific gravity of the urine sample with a urinometer. If a urinometer is not available, weigh a small beaker. Add 10 mL of the urine sample, and reweigh. Calculate the density and specific gravity of each sample.

B. UREA

Place 5 mL of each urine sample in separate test tubes. Add 2 mL 1% urease solution to each. Let the mixtures stand for 1 hour. Place a piece of moist, red litmus paper across the top of each test tube. If no change occurs in the color of the litmus paper, gently heat the test tubes for one or two minutes. If urea is present, the evolution of ammonia will turn the paper blue, indicating the presence of urea. Record results.

$$H_2O + \underset{\text{urea}}{NH_2\overset{\overset{\displaystyle O}{\|}}{C}NH_2} \xrightarrow{\text{urease}} \underset{\text{ammonia}}{2\ NH_3} + CO_2$$

C. ELECTROLYTES

C.1 *Chloride ion, Cl^-* Place 5 mL of each urine sample in separate test tubes. Add 10 drops of 6 M HNO_3, and 10 drops 0.1M $AgNO_3$. A white precipitate(AgCl) confirms the presence of chloride.

C.2 *Sulfate ion, SO_4^{2-}* Place 5 mL of each urine sample in separate test tubes. Add 10 drops of 6 M HNO_3, and 10 drops of 0.1 M $BaCl_2$. Place the test tubes in a warm water bath for 10 minutes. A white precipitate ($BaSO_4$) confirms the presence of sulfate.

C.3 *Phosphate ion, PO_4^{3-}* Place 5 mL of each urine sample in separate test tubes. Add 10 drops of 6 M HNO_3 and 2 mL ammonium molybdate solution. Heat gently. A yellow precipitate confirms the presence of phosphate.

C.4 *Sodium ion, Na^+* Clean a flame-test wire in 6 M HCl and then dip the wire in one of the urine specimens. Heat the liquid on the wire loop in a flame. A bright, yellow-orange flame indicates the presence of sodium ion.

D. BAR GRAPH OF ELECTROLYTES

On the graph paper, represent the data given as a bar graph of the electrolytes (cations and anions) typically found in body fluids.

E. GLUCOSE

Place 10 drops of each urine sample in separate test tubes. Add 5 mL Benedict's reagent to each. Place the test tubes in a boiling water bath for 5 minutes. Cool. Record any changes in color. If glucose is present, estimate the amount.

Color with Benedict's reagent	mg%	mg/dL
Blue	0.10	100
Blue-green	0.25	250
Green	0.50	500
Yellow	1.00	1000
Orange	2.00	2000

Commercially available test strips such as Clinistix or tablets such as Clinitest may also be used to test for the presence of glucose. Follow directions on the package. Report observations.

F. KETONE BODIES

Place 5 mL of each urine sample in separate test tubes. Saturate each with ammonium sulfate, $(NH_4)_2SO_4$. Add 5 drops of 5% nitroprusside reagent. Tip the test tube and *CAREFULLY* add 20 drops of concentrated NH_4OH down the side. The formation of a purple ring where the layers meet indicates the presence of ketone bodies, acetone and/or acetoacetic acid.

 USE EXTREME CARE IN HANDLING CONCENTRATED NH_4OH

Commercially available test strips such as Ketostix may also be used to test for the presence of ketone bodies. Follow directions on the package. Report results.

G. PROTEINS

Place 5 mL of each urine sample in separate test tubes. Heat the solution to boiling for one to two minutes. If a precipitate forms, add 5 drops of 1 M HAc (acetic acid). Heat for 1 more minute. The formation of a white cloudy precipitate indicates the presence of protein.

Commercially available test strips such as Albustix may also be used to determine the presence of protein (albumin) in the urine samples.

NAME _____ SECTION _____ DATE _____

EXPERIMENT 42
CHEMISTRY OF URINE
LABORATORY REPORT

A. COLOR, pH AND SPECIFIC GRAVITY

		Normal	Pathological
A.1	Color		
A.2	pH		
A.3	Specific gravity		

Alternate determination of specific gravity:

	Normal	Pathological
Mass of beaker + specimen		
Mass of beaker		
Mass of urine		
Density		

B. UREA

	Normal	Pathological
Color change		
Urea (yes/no)		

C. ELECTROLYTES

Indicate the presence of electrolytes as follows:
Not present (-)   Present(+)   Strongly Present (++)

		Normal	Pathological
C.1	Cl^-		
C.2	SO_4^{2-}		
C.3	PO_4^{3-}		
C.4	Na^+		

EXPERIMENT 42

D. BAR GRAPH OF ELECTROLYTES

The following lists the component in a typical urine sample at a pH of 5.5. Draw a bar graph to represent this data on the graph paper.

Concentration (mmole/L) of Electrolytes in Urine pH 5.5

Cations		Anions		Nonelectrolytes	
Na^+	170	Cl^-	175	Urea	300
K^+	90	HPO_4^{2-}	45	Creatinine	10
Ca^{2+}	2	SO_4^{2-}	60		
Mg^{2+}	3	Organic acids	45		
NH_4^+	60				

mmoles/L
↓

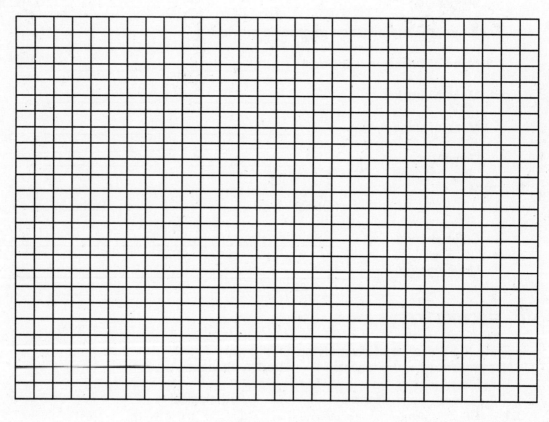

Ions

NAME_____ SECTION_____ DATE_____

E. GLUCOSE

	Normal	Pathological
Color with Benedict's reagent	_____	_____
% glucose	_____	_____
mg/dL	_____	_____

Optional

	Normal	Pathological
Test paper	_____	_____
Color	_____	_____
% glucose	_____	_____

When would glucose be found in a urine sample?

F. KETONE BODIES

	Normal	Pathological
Color	_____	_____
Ketone bodies?	_____	_____

Optional

	Normal	Pathological
Test paper	_____	_____
Results	_____	_____

When would ketone bodies be found in a urine sample? Why?

EXPERIMENT 42

G. PROTEIN

	Normal	Pathological
Appearance	_____	_____

Optional

Test paper	_____	_____
Results	_____	_____

When would protein appear in a urine sample?

EXPERIMENT 43
ENERGY PRODUCTION IN THE LIVING CELL

GOALS

1. Observe the reactions of glucose in glycolysis and fermentation.
2. Distinguish between aerobic and anaerobic reactions.

MATERIALS NEEDED

1 package dry yeast
fermentation tubes(5)
 (or 5 test tubes and
 5 small test tubes
 that fit inside)
10% glucose solution
10% sucrose solution
10% starch solution

250 mL Erlenmeyer flask
thermometer
methylene blue
mineral oil
temperature bath (37°C)
ethanol
beakers

CONCEPTS TO REVIEW

glycolysis
anerobic
aerobic
fermentation

BACKGROUND DISCUSSION

The amount of food you eat must be sufficient to meet all your metabolic needs. One of the most important requirements is the production of energy required by the cells to do work. Typically, energy is produced from glucose obtained from the carbohydrates in your diet. When the glucose is metabolized in the cells, a series of enzyme-catalyzed reactions extract energy from glucose and store it in ATP molecules.

Glycolysis

In human cells, glucose initially undergoes glycolysis, a series of reactions in which glucose is converted to pyruvic acid.

$$\text{Glucose} + 2\text{ NAD}^+ \longrightarrow 2\text{ Pyruvic acid} + 2\text{ NADH}$$

EXPERIMENT 43

Normally, when oxygen is available (aerobic conditions), hydrogens and electrons are transported to the electron transport system where another series of reactions occur that produce 2 or 3 ATP molecules. The hydrolysis of the high-energy bonds in ATP molecules provides energy to do work in the cell. The conversion of glucose to pyruvic acid under aerobic conditions can provide a total of 6 ATP.

$$\text{glucose} \longrightarrow 2 \text{ pyruvic acid} + 6 \text{ ATP}$$

We will observe the conversion of glucose to pyruvic acid using methylene blue as a hydrogen acceptor as well as an indicator of the reaction. The oxidized form of methylene blue is blue, while its reduced form is colorless.

$$\text{methylene blue (oxidized)} + 2H \longrightarrow \text{methylene blue (reduced)}$$
$$\text{blue} \qquad\qquad\qquad\qquad\qquad\qquad \text{colorless}$$

Under aerobic conditions, the pyruvic acid would be oxidized further to acetyl CoA and then to CO_2 and H_2O via the citric acid cycle. During these additional oxidative steps, NAD^+ and FAD carry hydrogens and electrons to the electron transport chain for additional ATP production. The overall conversion of glucose to CO_2 and H_2O provides a total of 36 ATP for the cell.

$$\text{glucose} + 6\, O_2 \xrightarrow{\text{aerobic}} 6CO_2 + 6H_2O + 36 \text{ ATP}$$

When cells do not have oxygen available (anaerobic conditions), the electron transport system does not operate. Then, pyruvic acid from glycolysis must be converted to lactic acid with no further oxidation possible. The anaerobic conversion of glucose to lactic acid provides only a small amount of energy for the cells. (This ATP is produced by direct phosphorylation only.)

$$\text{Glucose} \xrightarrow{\text{anaerobic}} \text{Lactic acid} + 2 \text{ ATP}$$

Fermentation

In cells of certain organisms such as yeast, there are enzymes that convert pyruvic acid to ethanol and CO_2, a process called *fermentation*. In this experiment, we will observe the fermentation reaction by observing the production of CO_2 bubbles in a mixture of glucose and yeast cells.

$$\text{Glucose} \xrightarrow{\text{yeast}} 2 \text{ Ethanol} + 2\, CO_2$$

LABORATORY ACTIVITIES

 Safety goggles are required!

A. FERMENTATION

Preparation of Yeast Suspension
Fill a large beaker about half full of water and warm to 37°C. Place 5 g of yeast in a 250 mL Erlenmeyer flask. Add 100 mL of distilled water and stir thoroughly. Place the flask in the warm water bath for 10 minutes to bring the temperature of the yeast suspension to 37°C. Do *not* overheat.

Pour 25 mL of the yeast suspension into a large test tube or 125 mL Erlenmeyer flask. Fill a 250 or 400 mL beaker about half full of water. Bring the water to a boil. Place the test tube or flask containing the 25 mL yeast suspension in the boiling water for at least 10 minutes. This will be the *boiled* yeast suspension.

Preparation of Fermentation Tubes
Place the following solutions in five separate fermentation tubes.

Test tube	Solutions
1	10 mL yeast suspension and fill with water
2	10 mL yeast suspension and fill with glucose solution
3	10 mL yeast suspension and fill with starch solution
4	10 mL boiled yeast and fill with glucose solution
5	10 mL yeast and fill with sucrose solution

Set the tubes in a 37°C water bath, tray, water trough or some beakers. Use warm water (37°C) to cover as much of the fermentation tubes as possible without causing them to tip over. See Figure 43.1.

Figure 43.1 Fermentation tube.

EXPERIMENT 43

Using test tubes as fermentation tubes

If fermentation tubes are not available, place the solutions in five separate test tubes. Place the smaller test tubes upside down in the large test tubes. Place your hand firmly over the mouth of the test tube and invert. When the small test tube inside has completely filled with the mixture, return the larger test tube to an upright position. See Figure 43.2. Place the tubes in a 37°C water bath and watch for the appearance of bubbles in the small test tubes.

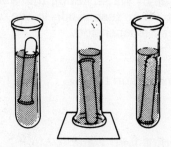

Figure 43.2 *Test tubes used as fermentation tubes.*

Observations of fermentation

Check the fermentation tubes every 30 min during the laboratory period. Add some warm water to the water bath, if necessary, to keep the temperature at about 37°C. Look for the formation of bubbles at the top of the closed tube. If the fermentation tubes have markings, record the volume of gas each time. If not, use a small ruler, to measure the height of the space in the tube occupied by the CO_2 bubble. See Figure 43.3.

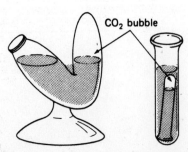

Figure 43.3 *Fermentation tubes with CO_2 bubble.*

At the end of the testing period, note the odor of each sample. Check the odor of a small sample of ethanol to see if ethanol has been produced.

B. GLYCOLYSIS (ANAEROBIC CONDITIONS)

Add 2 mL of methylene blue to three test tubes. Prepare the test tubes as follows:

Test Tube	Add
1	5 mL water and 3 mL yeast suspension
2	5 mL glucose and 3 mL yeast suspension
3	5 mL glucose and 3 mL boiled yeast suspension

Add mineral oil dropwise until a 2-3 mm layer forms. This will prevent the oxygen of the air from contact with the reaction mixture. Record the initial (0 minutes) color of the mixtures in the test tubes. Place the test tubes in a beaker of warm water (37°C). Every 30 minutes, observe the color of the mixtures. When methylene blue is reduced, its blue color fades. Record your observations.

EXPERIMENT 43
ENERGY PRODUCTION IN THE LIVING CELL
LABORATORY REPORT

A. FERMENTATION

Observations
Test tube

Time	1	2	3	4	5
0 minutes					
30 minutes					
60 minutes					
90 minutes					
120 minutes					

Describe any odor produced by the reaction.

_____

_____

1. In which test tubes did fermentation occur? Why?

2. In which test tube did fermentation not take place? Why?

3. Write the equation for the fermentation reaction in yeast cells.

EXPERIMENT 43

B. GLYCOLYSIS (ANAEROBIC CONDITIONS)

	Observations Test tube		
Time	1	2	3
0 minutes	_____	_____	_____
30 minutes	_____	_____	_____
60 minutes	_____	_____	_____
90 minutes	_____	_____	_____
120 minutes	_____	_____	_____

1. In which test tubes did glycolysis occur? Why?

2. In which test tubes did glycolysis not occur? Why?

3. What are the possible end products of pyruvic acid from glycolysis in aerobic conditions?

   How does this differ under anaerobic conditions?

APPENDIX A CONSTRUCTING A GRAPH

DATA TABLE

Let us suppose that we wanted to measure the distance and time that a bicycle ride traveled. We might take measurements of the distance covered by the rider at certain times. This would give a data table that might look like this.

Distance Covered by a Bicycle Rider with Time

DATA TABLE

Time (hr)	Distance Covered (km)
1	5
3	14
4	20
6	30
7	33
8	40
9	46
10	50

Now we are ready to construct a graph to represent this information.

CONSTRUCTION OF THE GRAPH

1. Draw vertical and horizontal axes on the graph paper. The lines should be set in to leave a margin for numbers and labels. The idea is to fully utilize the graph paper, not just a corner. Place a title at the top of the graph. The title should be derived from the data table. See Figure A.1.

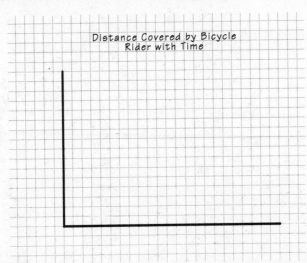

Figure A.1 *Vertical and horizontal axes drawn on graph paper.*

2. Label each axis with the type of measurement and its units. The labels and units are derived from the measurements you took when you prepared the data table. Labels in this graph would be distance (km) on the vertical axis and time(hr) on the horizontal axis. See Figure A.2.

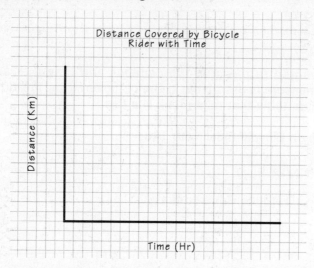

Figure A.2 *Labels written on each axis.*

3. Each axis represents a scale for that measurement. To select a proper scale, observe the lowest and highest numbers in each set of data. The values represented on an axis must be **equally spaced** and fit on the line you have drawn. The scale cannot exceed the line nor should it cover only a small portion of the line. Utilize as much of the full space as possible.

For our sample graph, we might set up a scale for distance that starts at 0 km and goes to 60 km. A convenient interval might be 5 or 10 km. Note that you only have to mark a few lines in order to interpret the scale. It gets too crowded with numbers to try and number every line. The time scale uses time intervals of 1 hour to cover the 10 hour time span for the experiment. See Figure A.3.

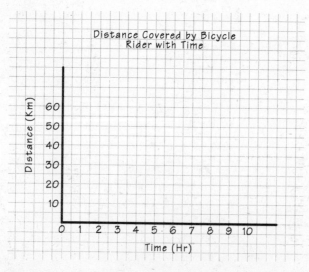

Figure A.3 *Marking equal intervals for data on each axis.*

PLOTTING THE INFORMATION

4. Now you are ready to plot the data on the graph. Each pair of measurements makes a point on the graph. For example, the rider has covered 20 km at a time of 4 hr. Find 20 km on the distance scale, and 4 hr on the time scale, and imagine perpendicular lines that intersect. That point of intersection is the plot of that set of measurements. Plotting all of the points will begin to show the relationship between distance and time. (You do not need to draw intersecting lines - just find the point of intersection.) Sometimes a small circle is drawn around the point to make it more visible.

5. **Connect** the points. However, this does not mean to jump from point to point in a zig zag fashion. It means drawing a smooth line or curve that goes through most of the points, but perhaps not all. Some points will fall off the line or curve you draw because there is some error associated with any measurement. Rarely will all points fall on a straight line. See Figure A.4.

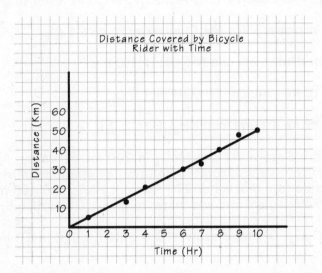

Figure A.4 *A completed graph with data points connected in a smooth line.*

APPENDIX B SOLUTIONS FOR THE LABORATORY

Acetic acid, 10%;, dilute 10 mL glacial HAc with water to 1.0 L

Acetic acid, 0.1M; dilute 6 mL glacial acetic acid with water to 1.0 L

Acetic acid, 1 M; dilute 60 mL glacial acetic acid with water to 1.0 L

Acetic acid, 6 M; dilute 350 mL glacial acetic acid with water to 1.0 L

Alanine, 1%; dissolve 1.0 g in water to give 100 mL

Albumin (egg) 1%; dissolve 10 g egg albumin in water to give 1.0 L

Aluminum sulfate, $Al_2(SO_4)_3$, 1%; dissolve 15 g $Al_2(SO_4)_3 \cdot 9H_2O$ in water to give 1.0 L

Ammonium chloride, 0.1 M; dissolve 5.4 g NH_4Cl in water to give 1.0 L

Ammonium hydroxide, 0.1 M; dilute mL con NH_4OH to 1.0 L

Ammonium hydroxide, 1 M; dilute mL con NH_4OH to 1.0 L

Ammonium hydroxide, 6 M; dilute 400 mL con NH_4OH to 1.0 L

Ammonium molybdate solution; dissolve 81 g H_2MoO_4 in 200 mL of water. Add 60 mL concentrated NH_4OH to give a saturated solution of ammonium molybdate. Filter. Slowly add filtrate to a mixture of 270 mL con. HNO_3 and 400 mL water. Let stand for 1 day. Filter again, and add water to give 1.0 liter of solution.

Ammonium oxalate, $(NH_4)_2C_2O_4$, 0.1 M; dissolve 12.6 g $(NH_4)_2C_2O_4 \cdot H_2O$ in water to give 1.0 L

Aspartic acid, 1%; dissolve 1.0 g in water to give 100 mL

Barium chloride; $BaCl_2$ 0.1 M; dissolve 24 g $BaCl_2 \cdot 2H_2O$ in water to give 1.0 L

Benedict's reagent; dissolve 173 g sodium citrate, $Na_3C_6H_5O_7$, and 100 g anhydrous sodium carbonate, Na_2CO_3, in 800 mL water. Warm. Filter. Dissolve 17.3 g of $CuSO_4 \cdot 5H_2O$ in 100 mL of water. Combine with the sodium citrate solution, stir, and add water to give 1.0 liter of solution.

Bromine in 1,1,1-trichloroethane, 2% solution; dissolve 2 mL Br_2 in 100 mL of CCl_3CH_3. Keep in hood!

Bromphenol blue; Add 0.1 g bromphenol blue to 7.5 mL of 0.02 M NaOH and dilute to 250 mL with distilled water

Butanol:glacial acetic acid: water (3:1:1) 300 mL butanol + 100 mL glacial acetic acid + 100 mL water

Calcium chloride, $CaCl_2$, 0.1 M; 14.7 g $CaCl_2 \cdot 2H_2O$ in water to give 1.0 L

Calcium chloride, $CaCl_2$, 5%; 66 g $CaCl_2 \cdot 2H_2O$ in water to give 1.0 L

Copper sulfate, $CuSO_4$, 0.5% ; dissolve 7.8 g $CuSO_4 \cdot 5H_2O$ in water to give 1.0 L

Copper sulfate, $CuSO_4$, 0.1 M ; dissolve 25 g $CuSO_4 \cdot 5H_2O$ in water to give 1.0 L

Cysteine, 1%; dissolve 1 g cysteine in water to give 100 mL

Fructose, 1%; dissolve 1 g fructose in water to give 100 mL

Gelatin, 1%; dissolve 10 g gelatin in water to give 1.0 L

Glucose, $C_6H_{12}O_6$, 1%; dissolve 1 g glucose in water to give 100 mL

Glucose, $C_6H_{12}O_6$, 5%; dissolve 50 g glucose in water to give 1.0 L

Glucose, $C_6H_{12}O_6$, 10%; dissolve 100 g glucose in water to give 1.0 L

Glucose, $C_6H_{12}O_6$, 1 M; dissolve 180 g glucose in water to give 1.0 L

Glutamic acid, 1%; dissolve 1 g in water to give 100 mL

Glycine, 1%; dissolve 1 g glycine in water to give 100 mL

Hard water; magnesium sulfate; dissolve 10 g $MgSO_4$ in water to give 1.0 L

Hydrochloric acid, 10%; dilute 100 mL con. HCl to 1.0 L

Hydrochloric acid, 0.1 M; dilute 8 mL of con. HCl to 1.0 L

Hydrochloric acid, 1 M; dilute 83 mL of con. HCl to 1.0 L

Hydrochloric acid, 6 M; Dilute 500 mL con. HCl to 1.0 L

Hydrochloric acid, con.; Use concentrated HCl

Iodine reagent, I_2/KI; dissolve 10 g I_2 + 20 g KI in water to give 1.0 L

Iodine solution (Vitamin C titration); dissolve 16 g KI and 1.5 g I_2 in water to give 1.0 L

Iron (III) chloride, $FeCl_3$, 1%; dissolve 17 g $FeCl_3 \cdot 6 H_2O$ in water to give 1.0 L

Iron (III) chloride, $FeCl_3$, 5%; dissolve 85 g $FeCl_3 \cdot 6 H_2O$ in water to give 1.0 L

Iron (III) chloride, $FeCl_3$, 0.1 M; Dissolve 27 g $FeCl_3 \cdot 6H_2O$ in water to give 1.0 L

Lactose, 1%; dissolve 1 g lactose in water to give 100 mL

Lead acetate 1%; dissolve 12 g $Pb(C_2H_3O_2)_2 \cdot 3H_2O$ in water to give 1.0 L

Lysine, 1%; dissolve 1 g in water to give 100 mL

Magnesium chloride, 5%; dissolve 50 g $MgCl_2$ in water to give 1.0 L

Maltose, 1%; dissolve 1 g maltose in water to give 100 mL

Ninhydrin, 0.2%, dissolve 0.2 g ninhydrin in 100 mL of 1-butanol saturated with water. Place in an aspirator for use. Commercial spray can of ninhydrin are also available.

Nitric acid, 10%, dissolve 10 mL HNO_3 in water to give 1.0 L

Nitric acid, HNO_3, 6 M, dilute 375 con HNO_3 to 1.0 L

Nitroprusside reagent, 5%; 10 g sodium nitroprusside in 90 mL water. Add 1 mL con. HNO_3

Pancreatin solution, 2%; dissolve 2 g pancreatin in 100 mL 0.25% sodium carbonate. Use immediately.

Pepsin, 2%; add 2 g of pepsin to 100 mL water.

Phenylalanine, 1%; dissolve 1 g in water to give 100 mL

Phenolpthalein 1%; dissolve 1 g phenolpthalein in 50 ml ethanol and 50 mL water.

Potassium chloride, KCl, 0.1 M; dissolve 7.5 g KCl in water to give 1.0 L

Potassium dichromate/H_2SO_4, 2% $K_2Cr_2O_7$, dissolve 20 g $K_2CR_2O_7$ in 100 mL 6 M H_2SO_4. Dilute carefully to 1.0 L

Potassium permanganate, $KMnO_4$ 0.1 M; dissolve 15.8 g $KMnO_4$ in water to give 1.0 L

Potassium thiocyanate, KSCN, 0.1 M; dissolve 9.7 g KSCN in water to give 1.0 L

Saliwanoff's reagent; dissolve 1.5 g of resorinol in 1 L of 6 M HCl

Serine, 1%; dissolve 1 g in water to give 100 mL

Silver nitrate, $AgNO_3$, 1%; dissolve 10 g $AgNO_3$ in water to give 1.0 L. Store in a dark bottle.

Silver nitrate, $AgNO_3$, 0.1 M; dissolve 17 g $AgNO_3$ in water to give 1.0 L. Store in a dark bottle.

Sodium carbonate, Na_2CO_3, 0.1 M; dissolve 23.2g $Na_2CO_3 \cdot 7H_2O$ in water to give 1.0 L

Sodium carbonate, Na_2CO_3, 1 M; dissolve 232 g $Na_2CO_3 \cdot 7H_2O$ in water to give 1.0 L

Sodium chloride, NaCl, 1%; dissolve 10 g in water to give 1.0 L

Sodium chloride, NaCl, 10%; dissolve 100 g in water to give 1.0 L

Sodium chloride, NaCl, 0.1 M; dissolve 5.8 g NaCl to water to give 1.0 L

Sodium chloride, NaCl, saturated; add 500 g NaCl to water to give 1.0 L

Sodium hydroxide, NaOH, 10%; dissolve 100 g NaOH in water to give 1.0 L

Sodium hydroxide, NaOH, 20%; dissolve 200 g NaOH in water to give 1.0 L

Sodium hydroxide, NaOH, 0.1 M; dissolve 240 g NaOH in water to give 1.0 L

Sodium hydroxide, NaOH, 6 M; dissolve 240 g NaOH in water to give 1.0 L

Sodium phosphate, 0.1 M; dissolve 38.0 g of $Na_3PO_4 \cdot 12\ H_2O$

Sodium sulfate, 0.1 M; dissolve 32 g $Na_3SO_4 \cdot 10H_2O$ in water to give 1.0 L

Starch, 1%; Make a paste of 10 g soluble starch. Slowly add 200 mL of boiling water with stirring. Add water to 1.0 L. The next day, pour off top layer of starch solution. If it is to be stored, add 0.01 g HgI_2.

Starch, 10%; Make a paste of 100 g soluble starch. Slowly add 200 mL of boiling water with stirring. Add water to 1.0 L. The next day, pour off top layer of starch solution. If it is to be stored, add 0.01 g HgI_2.

Sucrose, $C_{12}H_{22}O_{11}$, 10%; dissolve 100 g in water to give 1.0 L

Sucrose, $C_{12}H_{22}O_{11}$, 1 M; dissolve 342 g in water to give 1.0 L

Tyrosine, 1%; dissolve 1 g tyrosine in water to give 100 mL

Urease, 1%; dissolve 1 g urease in water to give 100 mL; use same day

Urine specimen (normal); mix 2 mL of 0.1 M NaCl, 2 mL of 0.1 M Na_2SO_4, 2 mL of 0.1 M Na_3PO_4. Add 1 g of urea. Adjust pH to 5.5-7.0 with 5-10 drops of 6 M HCl and dilute to 100 mL.

Urine specimen (pathological); To 100 mL of the normal urine specimen (above), add 2 g glucose, 2 g egg albumin, 4 mL acetone, and 1 mL 6 M HCl.

Atomic Numbers and Weights of the Elements

name	symbol	atomic number	atomic weight	name	symbol	atomic number	atomic weight
actinium	Ac	89	(227.0)	neodymium	Nd	60	144.2
aluminum	Al	13	27.0	neon	Ne	10	20.2
americium	Am	95	(243)	neptunium	Np	93	237.0
antimony	Sb	51	121.8	nickel	Ni	28	58.7
argon	Ar	18	39.9	niobium	Nb	41	92.9
arsenic	As	33	74.9	nitrogen	N	7	14.0
astatine	At	85	(210)	nobelium	No	102	(259)
barium	Ba	56	137.3	osmium	Os	76	190.2
berkelium	Bk	97	(247)	oxygen	O	8	16.0
beryllium	Be	4	9.0	palladium	Pd	46	106.4
bismuth	Bi	83	209.0	phosphorus	P	15	31.0
boron	B	5	10.8	platinum	Pt	78	195.1
bromine	Br	35	79.9	plutonium	Pu	94	(242)
cadmium	Cd	48	112.4	polonium	Po	84	(210)
calcium	Ca	20	40.1	potassium	K	19	39.1
californium	Cf	98	(251)	praseodymium	Pr	59	140.9
carbon	C	6	12.0	promethium	Pm	61	(145)
cerium	Ce	58	140.1	protactinium	Pa	91	(231.0)
cesium	Cs	55	132.9	radium	Ra	88	(226.0)
chlorine	Cl	17	35.5	radon	Rn	86	(222)
chromium	Cr	24	52.0	rhenium	Re	75	186.2
cobalt	Co	27	58.9	rhodium	Rh	45	102.9
copper	Cu	29	63.5	rubidium	Rb	37	85.5
curium	Cm	96	(247)	ruthenium	Ru	44	101.1
dysprosium	Dy	66	162.5	samarium	Sm	62	150.4
einsteinium	Es	99	(252)	scandium	Sc	21	45.0
erbium	Er	68	167.3	selenium	Se	34	79.0
europium	Eu	63	152.0	silicon	Si	14	28.1
fermium	Fm	100	(257)	silver	Ag	47	107.9
fluorine	F	9	19.0	sodium	Na	11	23.0
francium	Fr	87	(223)	strontium	Sr	38	87.6
gadolinium	Gd	64	157.3	sulfur	S	16	32.1
gallium	Ga	31	69.7	tantalum	Ta	73	180.9
germanium	Ge	32	72.6	technetium	Tc	43	(99)
gold	Au	79	197.0	tellurium	Te	52	127.6
hafnium	Hf	72	178.5	terbium	Tb	65	158.9
helium	He	2	4.0	thallium	Tl	81	204.4
holmium	Ho	67	164.9	thorium	Th	90	232.0
hydrogen	H	1	1.0	thulium	Tm	69	168.9
indium	In	49	114.8	tin	Sn	50	118.7
iodine	I	53	126.9	titanium	Ti	22	47.9
iridium	Ir	77	192.2	tungsten	W	74	183.9
iron	Fe	26	55.8	unnilennium	Une	109	(266)
krypton	Kr	36	83.8	unnilhexium	Unh	106	(263)
lanthanum	La	57	138.9	unnilpentium	Unp	105	(262)
lawrencium	Lr	103	(260)	unnilquadium	Unq	104	(261)
lead	Pb	82	207.2	unnilseptium	Uns	107	(262)
lithium	Li	3	6.9	uranium	U	92	238.0
lutetium	Lu	71	175.0	vanadium	V	23	50.9
magnesium	Mg	12	24.3	xenon	Xe	54	131.3
manganese	Mn	25	54.9	ytterbium	Yb	70	173.0
mendelevium	Md	101	(258)	yttrium	Y	39	88.9
mercury	Hg	80	200.6	zinc	Zn	30	65.4
molybdenum	Mo	42	95.9	zirconium	Zr	40	91.2

*Parentheses indicate the atomic weight of the most stable best-known isotope.

Functional Groups in Organic Molecules

Class	Functional group	Example	Name (ending)
alkane	none (C—C bonds)	CH$_3$CH$_3$	eth**ane**
alkyl halide	F, Cl, Br, or I halogen atom	CH$_3$CH$_2$Cl	chloroethane (**ethyl chloride**)
alkene	$\diagdown$C=C$\diagup$ double bond	H$_2$C=CH$_2$	eth**ene**
alkyne	—C≡C— triple bond	H—C≡C—H	eth**yne**
aromatic compounds	benzene ring	⬡	**benzene**
alcohol	—OH hydroxyl group	CH$_3$CH$_2$—OH	ethan**ol** (ethyl **alcohol**)
phenol	hydroxyl group on aromatic compound	OH–⬡	**phenol**
thiol	—SH group	CH$_3$CH$_2$—SH	ethane**thiol**
ether	oxygen atom between two alkyl groups	CH$_3$—O—CH$_3$	dimethyl **ether**
aldehyde	$\overset{O}{\underset{}{\overset{\|}{-C-H}}}$ carbonyl group	$\overset{O}{\underset{}{\overset{\|}{CH_3C-H}}}$	ethan**al** (acet**aldehyde**)
ketone	$\overset{O}{\underset{}{\overset{\|}{-C-}}}$ carbonyl group	$\overset{O}{\underset{}{\overset{\|}{CH_3-C-CH_3}}}$	propan**one** (dimethyl **ketone**)
carboxylic acid	$\overset{O}{\underset{}{\overset{\|}{-C-OH}}}$ carboxyl group	$\overset{O}{\underset{}{\overset{\|}{CH_3C-OH}}}$	ethan**oic acid** (ace**tic acid**)
ester	$\overset{O}{\underset{}{\overset{\|}{-C-O-}}}$	$\overset{O}{\underset{}{\overset{\|}{CH_3COCH_3}}}$	methyl ethan**oate** (methyl **acetate**)
amine	—N— amino group	CH$_3$—NH$_2$	methyl**amine**
amide	$\overset{O}{\underset{}{\overset{\|}{-C-N-}}}$	$\overset{O}{\underset{}{\overset{\|}{CH_3C-NH_2}}}$	ethan**amide** (acet**amide**)

Formulas and Names of Some Common Ions

Group number	Formula of ion	Name of ion	Group number	Formula of ion	Name of ion
1A	H^+ (aq)	hydrogen ion	7A	F^-	fluoride ion
	Li^+	lithium ion		Cl^-	chloride ion
	Na^+	sodium ion		Br^-	bromide ion
				I^-	iodide ion
2A	Mg^{2+}	magnesium ion			
	Ca^{2+}	calcium ion	Transition metals	Fe^{2+}	iron(II) ion
	Ba^{2+}	barium ion		Fe^{3+}	iron(III) ion
				Cu^+	copper(I) ion
3A	Al^{3+}	aluminum ion		Cu^{2+}	copper(II) ion
				Ag^+	silver ion
5A	N^{3-}	nitride ion		Zn^{2+}	zinc ion
	P^{3-}	phosphide ion			
6A	O^{2-}	oxide ion			
	S^{2-}	sulfide ion			

Formulas and Names of Some Common Polyatomic Ions

Charge	Formula of ion	Name of ion
1+	NH_4^+	ammonium ion
1−	OH^-	hydroxide ion
	NO_2^-	nitrite ion
	NO_3^-	nitrate ion
	HCO_3^-	hydrogen carbonate (bicarbonate) ion
	HSO_4^-	hydrogen sulfate (bisulfate) ion
	$C_2H_3O_2^-$	acetate ion
2−	CO_3^{2-}	carbonate ion
	SO_3^{2-}	sulfite ion
	SO_4^{2-}	sulfate ion
3−	PO_4^{3-}	phosphate ion

Periodic Table

Period number	Group 1A Alkali metals	2A Alkaline earth metals												3A	4A	5A	6A	7A Halogens	8A Noble gases
1	1 H 1.008																		2 He 4.003
2	3 Li 6.941	4 Be 9.012												5 B 10.81	6 C 12.01	7 N 14.01	8 O 16.00	9 F 19.00	10 Ne 20.18
3	11 Na 22.99	12 Mg 24.31				Transition elements								13 Al 26.98	14 Si 28.09	15 P 30.97	16 S 32.06	17 Cl 35.45	18 Ar 39.95
4	19 K 39.10	20 Ca 40.08	21 Sc 44.96	22 Ti 47.88	23 V 50.94	24 Cr 52.00	25 Mn 54.94	26 Fe 55.85	27 Co 58.93	28 Ni 58.69	29 Cu 63.55	30 Zn 65.38		31 Ga 69.72	32 Ge 72.59	33 As 74.92	34 Se 78.96	35 Br 79.90	36 Kr 83.80
5	37 Rb 85.47	38 Sr 87.62	39 Y 88.91	40 Zr 91.22	41 Nb 92.91	42 Mo 95.94	43 Tc (98)	44 Ru 101.1	45 Rh 102.9	46 Pd 106.4	47 Ag 107.9	48 Cd 112.4		49 In 114.8	50 Sn 118.7	51 Sb 121.8	52 Te 127.6	53 I 126.9	54 Xe 131.3
6	55 Cs 132.9	56 Ba 137.3	57* La 138.9	72 Hf 178.5	73 Ta 180.9	74 W 183.9	75 Re 186.2	76 Os 190.2	77 Ir 192.2	78 Pt 195.1	79 Au 197.0	80 Hg 200.6		81 Tl 204.4	82 Pb 207.2	83 Bi 209.0	84 Po (209)	85 At (210)	86 Rn (222)
7	87 Fr (223)	88 Ra 226	89† Ac (227)	104 Unq	105 Unp	106 Unh	107 Uns	108 Uno	109 Une										

Representative elements — Transition elements — Representative elements

Metals / Nonmetals

*lanthanides

58 Ce 140.1	59 Pr 140.9	60 Nd 144.2	61 Pm (145)	62 Sm 150.4	63 Eu 152.0	64 Gd 157.3	65 Tb 158.9	66 Dy 162.5	67 Ho 164.9	68 Er 167.3	69 Tm 168.9	70 Yb 173.0	71 Lu 175.0

†actinides

90 Th 232.0	91 Pa (231)	92 U 238.0	93 Np (237)	94 Pu (244)	95 Am (243)	96 Cm (247)	97 Bk (247)	98 Cf (251)	99 Es (252)	100 Fm (257)	101 Md (258)	102 No (259)	103 Lr (260)